# Buying Power

## THE BUSINESS OF DEFENSE IN AMERICA

## Wesley Sparks

FREILING
AGENCY

Published by Freiling Agency, LLC.

P.O. Box 1264
Warrenton, VA 20188

www.FreilingAgency.com

PB ISBN: 978-1-963701-77-7
E-book ISBN: 978-1-963701-78-4

# Table of Contents

# Acknowledgments

FOR SO MANY WHO DIRECTLY contributed to this effort:

My family, Alaine, Miles, Creighton, and Collins, thanks for giving me the space to chase this effort.

Thank you to the many friends and mentors who discussed these topics and contributed to this even if they didn't know:

Mike Wood, Nate Wike, Sean Quinn, Christina Battagliese, Joe Chan, Mike Milner, Stephen Hedger, Ross Coffman, Mark Bigelow, Darren Whiddon, Chris Conley, Cheryl Cotterell, Jared Jones, Caroline Jones, Steve Hedger, Patrick Malcor, Jon Milner, Matt Rasmussen, Robert Soto, Norm Wade, Jon Acuff, and Dave Bassett.

# Introduction

THE DEFENSE INDUSTRY IS NEVER "business as usual." All the functions across manufacturing, software, human resources, sales, and strategy exist. Then, they are multiplied by the uncertainty of Congress, changing Presidential administrations, and the actions of geopolitical rivals. The current framework of consolidation of large defense-only firms, insulation from commercial competition, and publicly traded companies contributing so fundamentally to the conduct of democracy is unique to the United States and unique to this era. The defense industry is doing business in "hard mode." In macroeconomics, a country gets the outcomes it incentivizes. Unfortunately, the incentives for a commercial enterprise to compete in this sector are extremely limited. Companies in private business enterprise would probably balk at the idea that their customer provides their own terms and conditions the *seller* must adhere to, and your customer makes their own laws limiting your potential profit margin.

The purpose of this book is to explain the history, politics, and market forces that have shaped the industry of warfare. Putting private industry in the central position of national security is a relatively unique American concept. Arguably, as far back as the days of Washington's quartermasters seeking gunsmiths, individual businesses have had a surprising amount of power and agency in creating a defense "industry." During the build-up of World War II, this unique American market of weapon and arms manufacturing, which was just part of the main-street commercial enterprise, skyrocketed and spread from coast to coast. There are government installations and company-owned defense article facilities across all 50 states for valid and, unfortunately too often, absurd reasons.

There are industry hubs based on common sense geography, and the shipyards luckily happen to be on coasts (a cynic may assume there is a Congressional representative in the central plains who would make the case for shipbuilding in their district). The military vehicle center of gravity is in the Detroit area, a legacy from the automotive engineering pedigree of Chrysler Defense and General Motors churning out tanks. Then we have hubs like Radford Army Ammunition Facility in Virginia that are left over from previous war efforts and are ostensibly contractor-managed. Still, it doesn't take much of an environmentalist to notice that we probably should not be mixing solvents, propellants, and other explosives on the banks of the New River.

The American system of allowing, promoting, enticing, cajoling, and even begging private enterprise to be involved in defense procurement may be our citizens' most significant interaction point with our national government after our annual IRS tax event. It would be hard to find a US citizen who doesn't work at a company that supplies products or services to the defense enterprise or doesn't own shares of defense companies in their 401k. This system differs significantly from almost all other NATO members and most of the First World. Many countries view defense products as not having commercial use and as a government responsibility. Ministries of Defense, as they are often called in Europe, will directly determine the items required and then use government foundries and facilities to manufacture them. Arguably, the US export of defense industry commercialization has been more significant than even our legendary Abrams tanks or Joint Strike Fighters.

Some believe that having giant bomb and weapon-manufacturing corporations in the S&P 500 is perverse. The other argument is that our market approach brings an often contentious topic out into the open and forces it to compete with ideas, products, and priorities beyond just a few hallways on military installations. The US defense budget is provided to the public at a reasonably granular level. Within a few minutes of searching, someone can determine how many light tactical trucks the US Army will procure in a given year and what company has the contract to provide them. Thousands of professionals across the government, military, and private industry are constantly debating and positioning for what should be in our national arsenal. The uniquely American approach at least puts this process, and its winners, out into the public sphere.

The Constitution's directive to "provide for the common defense" leaves wide lanes for interpretation of how much is enough. Do we as a nation need to provide and equip a force that can fight two wars concurrently as our National Military Strategy has often required? The policy decisions on when and where to apply American military power are often tilted and even directed by the hardware and software the industry can provide. A robust and responsive military and defense apparatus provides political leaders with space and resources to make decisions. This book does not call for a bigger military defense budget, or provide a how-to for getting contracts. This is an analysis of our apparatus, the historical events that shaped it, and the optimistic hope that the aspects of American industry involved in the defense business are *Buying Power*.

# CHAPTER 1

# The Economics of Warfare

*"...national wealth is increased and secured by national power, as national power is increased and secured by national wealth."*
—Friedrich List (1789-1846)

THROUGHOUT HISTORY, MILITARY STRENGTH HAS been a cornerstone of imperial power, national security, and geopolitical influence. The ability to field well-equipped armies and navies has often determined the rise and fall of civilizations. Military spending and the development of robust supply chains for military equipment have been critical issues for rulers and governments across the ages. The United States did not invent the concept of private industry serving the military goals of a nation, but the convergence of industrial production capacity, market-based capitalism, and globalization of trade in the 20th century created a uniquely American experiment for "buying power."

In ancient Greece and Rome, military strength was vital to state power. Greek city-states like Athens and Sparta allocated up to 50% of public funds to their militaries during conflicts. Athens invested heavily in its navy, establishing a large Mediterranean fleet to protect their trade routes.

For the Romans, military expenses comprised about 75% of the empire's budget at its peak, allowing them to maintain a standing army of around 450,000 troops. The mining, processing, and hand-crafting raw material into weapons at that scale is staggering. Skilled Greek armorers, known as "hoplopoioi" crafted bronze armor, while Rome transitioned from bronze to iron, utilizing materials from areas like Noricum (modern-day Austria). The creation of "fabricae", state-run arms factories across Roman territories, facilitated the mass production of standardized military equipment, crucial for Rome's dominance.

The Medieval times, including the Crusades, represented a massive mobilization of military resources. European monarchs and nobles invested heavily in equipment, supplies, and transportation for expeditions. This period also saw increased contact with advanced Middle Eastern metalworking techniques, leading to improvements in armor and weapon design.

The introduction of the blast furnace in Europe during the Late Middle Ages revolutionized iron production, allowing for the creation of higher-quality steel in larger quantities. This innovation would have extensive effects on military equipment manufacturing and the broader economy in the centuries to come. The period from 1500 to 1799 saw dramatic changes in military technology and production, driven by the age of exploration, the widespread adoption of gunpowder weapons, and the rise of standing national armies. Military spending during this era often consumed between 40% to 80% of national treasury expense, depending on whether the country was at peace or war.

As European nations embarked on voyages of exploration and colonization, the demand for sturdy, well-armed ships increased dramatically. Countries like England, Spain, Portugal, and the Netherlands invested heavily in their navies, requiring vast amounts of timber, iron, and copper.

The construction of a single first-rate 'ship of the line' could consume the equivalent of 50-60 acres of mature oak forest. This enormous demand led to the establishment of protected forests and the development of long-term timber management strategies. England's Royal Navy, for instance, became so concerned about future timber supplies that it initiated a major tree-planting program in the 1600s just for ship construction. The production of naval cannons became a critical industry. By the late 18th century, a typical 74-gun 'ship of the line' carried about 100 tons of guns, representing a significant investment in materials and manufacturing capacity. The ability to produce large numbers of high-quality cannons became a marker of a nation's industrial and military prowess.

On land, the widespread adoption of firearms led to the development of new industries and skills. The production of gunpowder itself became a strategic industry. By the 18th century, major powers were producing thousands of tons of gunpowder annually. France, for example, was producing over 2,000 tons of gunpowder per year by the 1700s. The establishment of standing armies in this period led to increased and more consistent military spending. Louis XIV of France maintained an army of over 300,000 men in peacetime, requiring enormous expenditures on equipment, supplies, and wages. This level of military investment was a key factor in France's dominance of European politics in the late 17th century.

The period from the Napoleonic Wars through the end of the 19th century saw a dramatic transformation in military manufacturing and spending, driven by the Industrial Revolution and the increasing scale of warfare. During the Napoleonic Wars, military spending reached unprecedented levels across the European continent. In Britain, military expenditures rose from 15 million pounds in 1792 to over 71 million pounds by 1814, consuming up to 80% of government spending. This massive investment allowed Britain to subsidize its allies and maintain the naval blockade that ultimately contributed to Napoleon's defeat.

Napoleon's Grande Armée required vast quantities of muskets, cannons, and other equipment. The French government established national arsenals and encouraged private manufacturers to increase production. By 1810, France was producing over 200,000 muskets annually, a testament to the scale of military-industrial production in this era.

As the 19th century progressed, the Industrial Revolution transformed military manufacturing. The introduction of rifled firearms in the mid-19th century and the adoption of breech-loading rifles spurred innovation in both design and production methods. These technological advancements required significant investment in new manufacturing equipment and processes. The American Civil War demonstrated the power of industrial might in warfare. The North's superior manufacturing capacity, particularly in firearms and artillery, played a crucial role in its victory. The war also saw advancements in the mass production of military equipment, with factories like the Springfield Armory producing hundreds of thousands of standardized rifles.

In Europe, the arms race between powers like Prussia, France, and Britain drove continuous innovation and increased military spending. Prussia's victory in the Franco-Prussian War was partly attributed to its superior artillery, produced by Krupp's steel works in Essen. (ThyssenKrupp continues as a major provider of industrial capacity for submarines and ships.) Naval technology advanced from wooden sailing ships to iron and then steel steam-powered vessels. The launch of HMS Warrior in 1860, the first iron-hulled, armored warship, marked a new era in naval construction and military spending. Its construction cost over £357,000 and was equivalent to about 1% of the entire British government budget for the year.

By the end of the 19th century, military-industrial complexes had emerged in many developed nations. In Germany, military spending increased from 90 million marks in 1872 to 438 million marks by 1912. The United States, despite a period of relative peace, increased annual military expenditures from about $20 million in 1870 to $170 million by 1900.

The history of military material sourcing and manufacture from ancient times to the late 19th century reflects not only technological advancements but also the critical role of military strength in shaping civilizations. Throughout this evolution, military spending has consistently been one of the largest expenditures, and the national power provided by strong militaries either secured peace or encouraged belligerence.

From the silver mines that funded the Athenian navy to the vast iron works of the Industrial Revolution, the ability to secure resources and transform them into military might has been a defining characteristic of powerful states and empires. The willingness and capacity to invest heavily in military production have often determined a nation's position on the global stage.

✦ ✦ ✦

## Start of the Modern Era

The United States Department of War was officially established in 1789, and has changed names and focus several times over the 200 years. Jared Diamond's concepts on geography and technology from the book *Guns, Germs, and Steel* would certainly contribute to understanding the development of our economy in relation to military strength. The transformation of the United States Department of War into the Department of Defense marked a significant shift in how America viewed its military apparatus and its role on the global stage. This change, occurring in the aftermath of World War II, reflected the evolving nature of international relations and the United States' emerging position as a superpower.

The pivotal moment came with the National Security Act of 1947. This comprehensive legislation, signed into law by President Harry S. Truman, aimed to reorganize the United

States' military and intelligence services to better meet the challenges of the post-war world. A key component of this act was the creation of the National Military Establishment, which would eventually become the Department of Defense.

The transition wasn't immediate. Initially, the National Military Establishment existed alongside the Department of War and the Department of the Navy. However, it quickly became apparent that a more unified structure was needed. In 1949, an amendment to the National Security Act transformed the National Military Establishment into the Department of Defense, finally replacing the Department of War.

This change was more than just a rebranding exercise. It represented a fundamental shift in how the United States approached its military strategy and national security. The term "Defense" was chosen deliberately, but the posture of US military power has been far more expeditionary and forward stationed around the world than isolationist. This aligned with the United States' assumed role as a global peacekeeper, with at least a stated intent to use "deterrence" and compelling realpolitik as an assurance for the status quo. Of course, maintaining status at the top of the global power chain requires frequent balancing.

The transformation also reflected the changing nature of warfare and national security threats. The Cold War era demanded a more holistic approach to defense, integrating military strategy, technological advancements, and the production capacities of private industry. The success of the Manhattan Project and the industrial might of mass production fundamentally changed the American approach to maintaining military capacity and exerting that power. Arguably, the Cold War division of the world into two halves required this shift in ideology. In order to achieve the growing national demand for force projection, the new department took on a broader mandate, overseeing not just traditional military operations but also cutting-edge technological innovations that would shape both military capabilities and private enterprise interaction between the economy, stock prices, and military capacity.

The quote at the start of the chapter launches one of the main considerations in analyzing the modern American approach to buying defense capacity. The German economist and political scientist Friedrich List raised this fundamental question: does an economy exist to protect and sustain a nation, or does a nation exist to protect and defend an economy? Today, the Department of Defense stands as one of the largest and most complex organizations in the world, embodying the breadth and depth of America's global military commitments. Its creation marked a pivotal moment in American history, signaling the nation's commitment to a leadership role on the world stage and its recognition of the changing nature of global security challenges. The shift from the Department of War to the Department of Defense represented more than a simple name change. It encapsulated a broader evolution in American military thinking, foreign policy, and national identity, reflecting the nation's journey from a young republic focused on territorial defense to a global superpower with worldwide responsibilities and interests.

❖ ❖ ❖

## Economic Ends to Justify the Means

*"...misleading to consider national defense merely in terms of tons of steel and man-hours of labor when the problem was political as well as economic"*
—Robert H. Connery

The unique American concept of placing private industry at the forefront of national security and defense, can be traced back to the days of Washington's quartermasters seeking gunsmiths. This approach, where individual businesses wield significant power and agency in creating a defense "industry," is a distinct feature of the American economic landscape. From the 1790s through 1940s, we had a messy oscillation between military and war production expansion and contraction. The American experiment extends to how we finance and build hardware and capacity for defense differently than almost all other nations in history. We are now in an era of balancing publicly traded companies (and their quarterly earnings reports) with global supply chains and great power competition.

Borrowing a term made famous by President Dwight Eisenhower, we can define the defense industry as the military-industrial complex, the network of companies, organizations, and government agencies involved in researching, developing, producing, and maintaining military equipment, technology, and services for national defense and security purposes. This co-dependent and interconnected web of competitors can broadly be called the Defense Industrial Base

(DIB). Many sectors, industries, and companies should arguably be considered part of the DIB but are not often thought of in the same thread as the large primes. The companies in nuclear power, the ever-growing space industry, commercial shipping, mining, computer chips, data centers, and compute power all sustain and contribute to the DIB.

All nation-states produce their military force through one of three loose dogmas. The first would be absolute governmental control of war materiel production. This would be easiest to define according to a monarchy that owns the forest that provides the lumber for ocean vessels, the shipyard, the workers that build the ships, and then, of course the control of navies at sea for their crown. A modern example could be the Soviet Ministry of Defense, which, by 1989, may have had the most internal control of staff, budget, and internal political power of the entire Commissariat compared with their other bureaus. This approach is definitely in contrast to the US approach, where individual companies choose to participate in the industry (or choose not to contribute).

The American system of involving private enterprises in defense procurement is not just a policy but a significant part of our citizens' lives. It may be citizens' most significant interaction point with our national government that remains levels removed from public perception. It would be hard to find a US citizen who doesn't work at a company that supplies products or services to the defense enterprise or doesn't own shares of defense companies in their 401k. In the 1990s, ESG investing (environmental, social, and governance) began to take root in major investment funds. It has often been used as a marketing tool to promote investment theses that align with the values of institutional investors.

In the 2020s many funds that may otherwise not be openly associated with defense firms have started to consider defense materiel under the ESG umbrella. This system of investor involvement in war production differs significantly from that of almost all other NATO members and most First World countries. Many countries view defense manufacture as a dirty business that should be kept out of public view. The US export of defense industry commercialization has been even more significant than our defense products. While the defense industry globally privatizes and commoditizes, there are still substantial differences in the American way of building up for war compared to allies and peer threats.

Giant bomb and weapon makers like Lockheed Martin, Boeing, and Northrop Grumman are traded in the S&P 500, and largely behave like other large conglomerates with quarterly earnings projections and stock price management. Critics call it perverse, defenders argue it drags a shadowy world into the spotlight. One aspect of this public nature is forcing the concept of American military power out in the marketplace of ideas, with defense stock prices serving as a vote of confidence in our foreign policy. Congress appropriated $892 billion for fiscal year 2025, but there have been public comments promising a number around $1 trillion. An amazing aspect is how openly available the details of this budget are.

The Constitution's vague nod to "common defense" fuels endless debate: How much is enough? Do we still need a force that can conduct full-scale war on two fronts, as past strategies demanded? Global tensions, conflicts, and security threats can significantly influence defense stock prices. Events such as the outbreak of war, escalating regional tensions, or terrorist attacks often lead to increased demand for defense products and services, driving up stock prices. Political decisions by the US government and responses to changing geopolitics can have an exponential impact on the value of publicly traded companies.

The dramatic swings of the new Trump administration between NATO abandonment of tariff trade wars, reducing the DOD by 8%, and then promising more than a 10% budget increase have all occurred in the first 90 days. A poignant critique of the sad connection between world events, politics, and economic benefits is the 1997 film *Wag the Dog*. As a political satire, it tells the story of a Washington, D.C. spin doctor, played by De Niro, who is hired by the White House to divert public attention from a scandal involving the President and an underage girl just days before the election. To accomplish this, a Hollywood producer is enlisted to fabricate a fake war with Albania. Together, they create a media frenzy around a supposed conflict, complete with patriotic songs, staged video footage, and even a "hero" played by Woody Harrelson, who is later revealed to be a mentally unstable convict.

❖ ❖ ❖

# From Policy to Products

*"Critics of national-defense policies often refer to the military-industrial complex as if it were a conspiracy imposed from above against the will and desires of the American people...but even Eisenhower actually suggested a subtler, more insidious meaning... the military-industrial complex is ourselves."*
—Nick Kotz, Wild Blue Yonder, 1988

The U.S. National Defense Strategy (NDS) process is a high-level effort to align the Pentagon's budget with global threats, culminating in the National Military Strategy (NMS) and shaping the Future Years Defense Program (FYDP), which is a five-year spending roadmap. Directed by the Secretary of Defense, the NDS sets broad priorities like countering China's navy or Russia's cyber threats, gathering inputs from the intelligence community, White House guidance, and global regional commanders called Combatant Commanders.

The process starts with a classified review led by the Office of the Secretary of Defense. Over 12 months, analysts, generals, and civilian leaders assess risks, from hypersonic missiles to terrorism. They consult allies, study adversaries' moves, and debate force readiness. The resulting NDS, released every four years (last in 2022, next in 2026), outlines goals: deter aggression, ensure lethality, and strengthen alliances. It's broad but pivotal, guiding all DOD plans.

The Joint Chiefs of Staff then craft the NMS, a classified document translating the NDS into military objectives. In six months, they detail force postures, like boosting Pacific air wings, and operational needs, such as AI for targeting. The NMS prioritizes capabilities, directly informing budget decisions. The FYDP, built by the Pentagon's comptroller, turns these strategies into dollars. For example, a 2022 NDS focus on China spurred $13 billion for Guam's defense in the 2025 FYDP. This cycle is supposed to ensure strategy drives investment, not the other way around. That alone is a uniquely American concept that contributes to monetary policy and incentivizes certain investments by industry.

The National Security Act of 1947 restructured the US military and intelligence agencies after World War II. It created the Department of Defense, the Joint Chiefs of Staff, and the unified combatant command structure. However, the Goldwater-Nichols Act of 1986 significantly strengthened the Combatant Commanders' role and authority. This legislation was enacted in response to operational challenges and inter-service rivalries that hindered the effectiveness of the US military. The act clarified the chain of command, giving Combatant Commanders direct authority over the forces assigned to their commands and making them responsible for the planning and executing military operations within their designated areas of global responsibility.

This organizing concept divides the globe into a chessboard, with Combatant Commanders as grandmasters moving pieces across regions like the Indo-Pacific or Middle East. The Goldwater-Nichols Act of 1986 set up this board, carving the world into zones where these commanders plan and act, whether it's countering China's navy or training allies in Europe. Each commander, like a coach calling plays, pinpoints what they need, stealth jets or cyber defenses, to achieve their mission.

The "operator" or action arm of this global strategy gets their vote through the Joint Capabilities Integration and Development System, or JCIDS. It is a kind of military wish list sorter. Commanders pitch needs, like a destroyer with hypersonic missiles, and JCIDS ranks them, ostensibly ensuring the Army, Navy, and Air Force align on priorities. Once approved, the Defense Acquisition System kicks in, a multi-step journey from sketch to battlefield. It begins with brainstorming concepts, testing technologies, building prototypes, and finally churning out weapons platforms.

◆ ◆ ◆

## The Arms Arm of Foreign Policy

*"WAR is a racket. It always has been. It is possibly the oldest, easily the most profitable, surely the most vicious. It is the only one international in scope. It is the only one in which the profits are reckoned in dollars and the losses in lives."*
—Smedley Butler, *War Is a Racket*

Policy and strategy are not made in a vacuum. While there are certainly windowless rooms with cubicle farms in the Pentagon that may feel soul-sucking, the administrators of our military response to the ever-shifting geopolitical reality do not have the luxury of using "Hope" as a method. Hoping that China will take a step back in the South China Sea or hoping that the Islamic State will not reconstitute and threaten entire nation-states. We have large corporations performing military-industrial functions because we are a capitalist country. Critics of American defense industrial policy should always balance their positions based on the reality of the alternatives. Handwringing and debating over whether we are a global hegemon can be a worthy thought experiment. Still, it is hard to imagine a scenario where China would exert global influence in a more nuanced or humane way than the United States.

The United States defense industry significantly shapes the ability to execute foreign policy and geopolitical strategies. As a large and influential sector in the American economy, the defense industry wields considerable power in international relations. Through its production of advanced weaponry, technology transfers, and military assistance programs, the defense industry is an essential tool for projecting US power and influence on a global scale.

The defense industry acts as an arm of US foreign policy by selling weapons and military equipment to foreign nations (when approved by the Department of State). The United States is the world's largest arms exporter, with American defense companies supplying a wide range of weaponry to allies and partners around the globe. These sales generate substantial revenue for the industry and help cement strategic relationships and alliances. The US shapes international dynamics by providing advanced military capabilities to friendly nations.

Transferring American military technology and expertise to foreign nations is a powerful geopolitical tool. By sharing cutting-edge weapons systems and technical know-how, the US can enhance the military capabilities of its allies and partners, making them more effective in combating shared threats and promoting common interests. This technology transfer also fosters long-term dependence on American military hardware and support, which helps to maintain US influence and leverage in critical regions.

The "Overton Window" has shifted several times since the end of World War II. There was a period right after the war when "American Exceptionalism" became the primary opinion, taught to school children. This broad acceptance meant that exporting the American way of life through force when necessary, was considered the default position. This was a significant generational shift from the previous two generations of isolationist public policy. Then several more changes (generally slowly accepted by both political parties) to support or cause regime changes. After September 11[th,] the idea of regime change or dismantling non-nation state actors became a major proponent of our foreign policy.

President Trump's shifting stance on Ukraine and NATO in 2025 has sent shockwaves through the American defense industry, threatening billions in the export market. His push to freeze military aid to Ukraine and demand NATO allies spend a much higher percentage of their Gross Domestic Product on defense has significantly changed the value proposition of American materiel. Trump's public clash with Ukraine's Volodymyr Zelenskyy, including halting billions in aid, signals a retreat from security guarantees that once clinched deals. Countries like Poland, Japan, and Australia are quickly shifting from reliable target importers of American defense products to local shoppers, looking for suppliers closer to home.

The defense industry's role as an arm of US foreign policy is subject to controversy. Critics argue that the industry's influence can lead to an overemphasis on military solutions to complex geopolitical challenges at the expense of diplomatic and economic engagement. The so-called "military-industrial complex," coined by President Dwight D. Eisenhower, refers to the powerful nexus between the defense industry, the military, and elected officials, which can create perverse incentives for perpetuating conflicts and military interventions. There are also concerns about the potential for arms sales to destabilize regions or enable human rights abuses, primarily when weapons are sold to authoritarian regimes or nations with a history of conflict.

Despite these criticisms, the US defense industry remains a central pillar of American foreign policy and geopolitical strategy. As the world shifts to a multi-polar security paradigm, the industry's contributions to US military power, technological superiority, and strategic partnerships are essential to protecting American interests and maintaining international stability.

❖ ❖ ❖

*"Military contingency planning and the accelerating rate of obsolescence of the tools of war increasingly served as stimulants to the national economy... [making it] the warfare state."*
—Historians Leslie Decker and Robert Seager II

The United States defense industry operates within a unique economic framework that sets it apart from its counterparts in other countries. This distinctiveness is rooted in a combination of factors, including the sheer scale of US defense spending, the close relationship between the government and private sector, the emphasis on technological innovation, and the global reach of American military power.

One of the US defense industry's most striking features is the enormous government spending on military hardware, research, and development. The United States consistently outspends all other nations on defense, with a budget that dwarfs its closest competitors. The Fiscal Year 2025 US defense budget stands at over $850 billion, more than the combined defense spending of the following ten countries. While directionally correct, a giant asterisk is needed for China, as they allocate their budgets across state-owned industries and companies, hiding defense expenditures as commercial investment. China has regularly been accused of severely under-reporting its actual defense spending. With their centrally managed economy, there is significant spending that should be accrued to defense policy acquisition and investment that is misattributed to other areas of their economy.

The close relationship between the US government and the private sector is another critical factor that sets the American defense industry apart. In the United States, defense companies work hand in hand with the Department of Defense and other government agencies to develop and produce military technologies. This collaboration is facilitated by a complex web of procurement policies and beneficial accounting procedures for research and development.

Another unique aspect of the US defense industry is its global reach and influence. American defense companies supply the US military and export significant weaponry and technology to allies and partners worldwide. The US government actively promotes these exports as a foreign policy tool, using arms sales to cement strategic relationships and enhance the military capabilities of friendly nations. This global market for American defense products creates additional economic opportunities for US companies while giving the United States significant leverage over other countries' defense policies and priorities.

The structure of the US defense industry is also distinct from that of many other countries. In the United States, the defense sector is dominated by a handful of large, diversified corporations, such as Lockheed Martin, Boeing, and Raytheon. These companies have the scale and resources to undertake massive, complex weapons programs and weather government spending fluctuations. At the same time, an entire ecosystem of smaller, specialized defense firms provide niche technologies and services.

One of the most critical gaps in the current American way of war and developing platforms, is the focus on large quantities of exquisite platforms and capital ships. Current conflicts are showing a pivot to expendable hardware like drones. The existential question is who moves the needle first to address this gap. Is the burden on industry to analyze warfare and propose the new kinds of assets it will require, or does the DOD need to refine its requirements and ask the industry to follow.

Despite these challenges, the US defense industry remains a powerful economic force domestically and globally. Its unique combination of government support, private sector innovation, and global influence sets it apart from defense industries in other countries. As the United States navigates an increasingly complex and competitive international security environment, its defense industry's economic strengths and weaknesses will continue to shape its military capabilities and foreign policy options.

❖ ❖ ❖

## Government Functions

*"Politics is the entertainment division of the military-industrial complex."*                —Frank Zappa

The concept of "inherently governmental functions" is crucial in determining the roles and responsibilities that should be performed exclusively by the US government and those that can be outsourced to private industry. The Federal Activities Inventory Reform (FAIR) Act of 1998 and the Office of Management and Budget (OMB) Circular A-76 provide guidance, but administrations approach the topic differently.

According to the FAIR Act, an inherently governmental function is "a function that is so intimately related to the public interest as to require performance by Federal Government employees." These functions involve the exercise of sovereign powers, such as the interpretation and execution of US laws, the command of military forces, the conduct of foreign relations, and the determination of agency policy. The act requires federal agencies to annually inventory their activities and classify them as inherently governmental or potentially non-governmental.

OMB Circular A-76, first issued in 1966 and revised several times, sets forth the policies and procedures for determining whether commercial activities should be performed by the government or outsourced to the private sector. The circular aims to promote competition and choice in the delivery of government services while also ensuring that federal employees perform inherently governmental functions.

The US government should be striving to avoid competing with private industry by outsourcing commercial activities whenever possible. This approach is based on the belief that the private sector is generally more efficient and cost-effective in delivering goods and services. By contracting out commercial functions, the government can focus its resources on core responsibilities and reduce the size and cost of the federal workforce. This concept is being tested by the conflicting approaches of the new administration. Are consulting and IT contracts inherently wasteful, or is offloading commercial responsibilities to the private sector a competitive advantage?

In addition to the FAIR Act and OMB Circular A-76, other laws and regulations govern the relationship between the government and private industry. For example, the Competition in Contracting Act of 1984 requires federal agencies to use competitive procedures when procuring goods and services from the private sector unless certain exceptions apply. The Federal Acquisition Regulation

(FAR) provides a comprehensive set of rules and guidelines for government procurement, including provisions related to inherently governmental functions and the use of contractors.

Despite these efforts to delineate the roles of government and private industry, the boundary between inherently governmental and commercial functions can sometimes be blurry. Agencies must exercise judgment in classifying their activities and ensuring that outsourcing decisions do not compromise the government's core responsibilities or the public interest. This topic is central to the concept of a private defense industry.

The corporate, free-market participation of the US defense industry significantly protects the American political environment, offering a unique set of advantages that help to insulate the sector from political volatility and ensure its long-term stability. This protection is rooted in the defense industry's deep integration with the private sector, its significant employment and economic growth source, and its strong relationships with elected officials and policymakers.

One of the critical ways the free-market structure of the US defense industry serves as a political protection is by distributing the benefits of defense spending across a wide range of companies, communities, and regions. Unlike in countries where the defense sector is primarily state-owned, the US defense industry comprises diverse private companies, ranging from large multinational corporations to small, specialized firms. This decentralized structure ensures that the economic benefits of defense spending are spread throughout the country, creating a broad base of stakeholders with a vested interest in maintaining a strong and well-funded military.

The geographic dispersion of the US defense industry also plays a crucial role in its political protection. Defense companies and suppliers are in every state, providing jobs and driving economic growth in all Congressional districts. This widespread economic impact creates a powerful incentive for elected officials to support defense spending and protect the industry's interests, regardless of their political party or ideology. Lawmakers who seek to cut defense budgets or cancel primary weapons programs often face strong opposition from their constituents, who rely on the defense industry for their livelihoods.

# CHAPTER 2

# The Biggest Customer

## *How a Monopsony Spends $1 trillion*

*"In the year 2054, the entire defense budget will purchase just one aircraft. This aircraft will have to be shared by the Air Force and Navy 3-1/2 days each per week except for leap year when it will be made available to the Marines for the extra day."*
—Norman Augustine, former Undersecretary
of the Army, 1984

THE UNITED STATES DEPARTMENT OF Defense is not invisible in the global economy, boasting an annual budget that dwarfs many nations' GDPs. The thing that makes the DOD unique is its status as a monopsony, the economic term for a single buyer in a market who wields immense purchasing power. Beyond the large budget, it steers additional investment in sectors that aren't directly counted in defense spending. With a budget nearing $1 trillion, the DOD can be described as the world's biggest customer. In reality, the DOD is a feudal confederation of competing interests and zero-sum budget stand-offs. That description was fitting even before the upending of the start of the second Trump administration. Since World War II, the United States' approach to largely privatized defense development and production capability has been relatively unique. There are many critics arguing the defense acquisition system has grown too much into a centralized, controlled framework not unlike the last days of the Soviet empire.

The sheer volume and diversity of DOD consumption is staggering. The Department buys everything from sophisticated military technology like fighter jets to everyday items like peanut butter and printer paper. As a monopsony, their purchasing decisions have significant ripple effects. Large franchise programs like the F-35 Joint Strike Fighter from Lockheed Martin can bend, shape, and distort entire supply chains of parts and skilled labor.

The DOD is not monolithic; the fights on budget levels and who gets funding lines cut across all Congressional districts and the Pentagon knife fights often escape the five-sided puzzle palace and make the industry and financial press. Budget allocation within the DOD is complex, divided across personnel salaries, operations and maintenance, procurement, and research and development. Each of these categories further subdivides into countless line items, making the Department's spending an intricate exercise in resource management. However, the monopsony status allows the DOD to shape markets, influence prices, and determine which products and services are produced. This power has far-reaching implications for suppliers, who must adapt to the DOD's demands to stay competitive. Large corporations that operate in the defense ecosystem have the challenge of bending the Congressional appropriation process into the structure of quarterly earnings reports for their investors.

❖ ❖ ❖

## Different Flavors of Customer

There are different "flavors" of customers across the DOD. There are customers with a wish list (requirements), customers with decision authority (acquisition officials), and customers with money (appropriators and budget officials). The intersection of political policy and the ability to perform procurement tasks occurs through the Service Secretaries, their Assistants, and Undersecretaries. The Senior Acquisition Executives (SAEs) play a crucial role in shaping the future of the Department. These civilian appointees oversee the acquisition of weapons systems and manage multi-billion-dollar budgets. Their positions require Senate confirmation and grant them significant authority, but their salaries are far lower than similar positions in private industry.

The list of political appointee roles is published every four years in a paperback volume with a plum-colored cover, hence the name Plum Book. There are many Constitutional scholars that disagree over the intent of the term "advice and consent of the Senate," but key positions like Secretary of Defense, Secretaries of the Services, and their chief Undersecretaries have so far always required a simple majority vote in the Senate to become "Senate confirmed." That process of selecting, nominating, and then getting confirmed is a blend of competence, political gamesmanship, and now a preference for inexperience, flavored as outsiders to "drain the swamp."

At its core, the role of the Senior Acquisition Executive revolves around leadership in military procurement. These individuals are responsible for guiding the acquisition process within their respective military branches, ensuring that the armed forces have the necessary equipment and technology to conduct their role in the National Military Strategy. The legal foundation for these positions stems from the Goldwater-Nichols Department of Defense Reorganization Act of 1986, which aimed to enhance inter-service cooperation and streamline the acquisition process.

Senior Acquisition Executives operate at the crossroads of military strategy, technological innovation, and fiscal management. They must strike a balance between the need for advanced capabilities and budget limitations. Additionally, they navigate complex relationships with defense contractors and work with Congress to secure funding for essential programs. The choices they make can significantly impact national defense strategy and the preparedness of the U.S. armed forces.

Historically, the qualifications for these appointees have been quite stringent. Candidates must demonstrate a deep understanding of defense acquisition processes, a successful track record in program management, and the ability to navigate the complex landscape of military procurement. While some criticize a "Revolving Door" appearance, there are only two real sources of qualified acquisition leaders: people with experience in the DOD or leaders from the defense industry.

❖ ❖ ❖

# The Sherpas

Below the Service-level Assistant Secretary role, the place where the rubber meets the road, are the Program Executive Officers (PEOs) and their Program Managers (PMs). PEOs serve as senior acquisition executives for collections of programs. Their role has grown in importance as defense acquisition has become more complex. PEOs are typically one or two-star Generals or the equivalent Senior Executive Service civilians and they lead 75 different offices across all the services.

Supported by a team of experts in engineering, logistics, contracting, and finance, PEOs adopt a portfolio management approach to balance resources across programs. Their primary function is really to execute "Programs of Record." These are the large multi-year programs that are the industrial scale model the DOD uses to field equipment (or software) widely across a service.

Under the PEOs, O-6 Colonels (or Captains in Navy terms) serve as Program Managers. These are the day-to-day leaders of individual acquisition programs. PMs are responsible for all aspects of their assigned programs, including cost, schedule, and technical performance. They must navigate a complex web of stakeholders, including military operators, industry partners, congressional overseers, and the public. There are certainly major elements of overlap with the civilian field of project management, but the military officers and government civilians staffing these offices have their own educational entities. The Defense Acquisition University (DAU) and the Dwight D. Eisenhower School for National Security and Resource Strategy (formerly the Industrial College of the Armed Forces, ICAF) provide the official training and certification for DOD personnel. Each institution has distinct origins and complementary missions focused on developing expertise in defense acquisition and resource management.

DAU, established in 1991 by DOD Directive 5000.57, evolved from a consortium of training commands into a centralized corporate university headquartered at Fort Belvoir, Virginia. Its creation responded to the Defense Acquisition Workforce Improvement Act (DAWIA), which mandated certification for acquisition personnel. DAU provides standardized training in acquisition, technology, and logistics for military, civilian, and federal contractor personnel. It offers courses across seven functional areas, like program management and contracting. Accredited by the American Council on Education, DAU delivers instructor-led, virtual, and online training, serving over 150,000 workforce members annually. Its curriculum, guided by best practices and industry input, ensures professionals master the DOD's complex acquisition processes.

The Eisenhower School, founded in 1924 as the Army Industrial College, became the ICAF in 1946 and was renamed in 2012 to honor Dwight D. Eisenhower, an alumnus. Part of the National Defense University, it prepares senior military and civilian leaders for strategic roles in national security resource management. Its 10-month program, culminating in a Master of Science in National Resource Strategy, emphasizes materiel acquisition, joint logistics, and industry studies. The Senior Acquisition Course, offered with DAU, trains professionals for senior acquisition roles through a rigorous curriculum including electives on policy and innovation.

The Program Managers trained by these academic institutions face numerous challenges in executing their responsibilities. They must balance often competing demands from various stakeholders, manage complex technical risks, and navigate the intricacies of the defense acquisition system. Successful PMs possess a rare combination of technical expertise, leadership skills, and business acumen. Then Major General Dave Bassett loved the example of the "Sherpa" for the role of the Acquisition career field. The Sherpa doesn't pick the mountain to climb, but they rely on experience to help the tourist climber, explaining where to avoid the dead bodies along the path. There is an almost constant culture battle with the "Operator" types believing the "salad-eating office" types in acquisition just don't have enough warfighter focus to get them the right hardware. Acquisition and recruiting are the two areas in the military directly interacts with private individuals and enterprises and arguably does not have a great track record in either.

❖ ❖ ❖

## Contracting Officers

The position of Contracting Officer (abbreviated to KO in the Army and still CO in other services) constitutes a linchpin in the federal acquisition framework. This role entails the administration of contracts for goods, services, and construction, ensuring compliance with statutory and regulatory mandates. Contracting Officers serve as the sole representatives authorized to commit the United States Government to contractual obligations, a function that actually facilitates military operations and infrastructure development. Their responsibilities cover the solicitation of proposals, negotiation of terms, awarding of contracts, oversight of performance, even ending in the termination of agreements when necessary. This process demands accurate adherence to legal standards and DOD-specific policies. The best KOs know when to strike clauses and when certain terms and conditions provide the best chance for program success.

The authority of Contracting Officers is a real legal and regulatory structure, these are not part-time seasonal assignments for other career fields to sprinkle in. Title 10, Section 2304 of the U.S. Code establishes the statutory basis for DOD procurement, authorizing contracts to meet operational needs. Complementing this, Title 41, Chapter 11 of the U.S. Code, known as the Federal Property and Administrative Services Act, provides a broader framework for federal acquisitions. The Federal Acquisition Regulation (FAR) operationalizes these statutes, with FAR 1.602 delineating the Contracting Officer's capacity to enter, administer, and terminate contracts, contingent upon acting within their delegated authority. FAR 1.601 stipulates that only individuals appointed via a formal warrant may bind the government. The Defense Federal Acquisition Regulation Supplement (DFARS), notably DFARS 201.602, refines these guidelines for DOD contexts. The warrant, issued by a senior official such as a Head of Contracting Activity, specifies the scope of authority, ranging from limited monetary thresholds to unlimited jurisdiction.

The significance of this role cannot be overstated. Contracting Officers manage expenditures totaling billions annually, safeguarding fiscal integrity while enabling military readiness. Their oversight ensures the timely acquisition of critical resources, from advanced weaponry to logistical support, thereby sustaining operational capabilities. They mitigate legal, financial, and performance risks, maintaining accountability in a complex procurement environment. Their decisions shape long-term strategic outcomes, such as multi-year contracts that influence industrial partnerships and defense planning.

Emerging proposals, such as the Department of Government Efficiency (DOGE) initiative introduced in 2025, raise major questions about who has the power to bind the US Government with contracts and who has the authority to terminate contracts. DOGE, associated with advocates of streamlined governance, seeks to reduce bureaucratic overhead and enhance efficiency. Replacing Contracting Officers with statutory authority presents formidable challenges. The specific training and years of practical experience equip KOs to handle high-value, intricate contracts. This proficiency is not easily replicated.

Workforce reductions are constricting the pool of qualified candidates, intensifying succession challenges as experienced officers retire. Some estimates say 30% of the DOD workforce is already eligible for retirement. While smaller contracts might be delegated to junior personnel or automated systems, high-stakes agreements, such as those for major defense systems, demand the seasoned insight of unlimited warrant holders. The DOD's procurement apparatus relies on the Contracting Officer's blend of legal authority, technical skill, and strategic foresight.

To put acquisition executives, program managers, and contracting officers into a football metaphor they each have different roles in a franchise. The acquisition executives are the General Managers and organization Presidents; they set policy and give broad direction. The program managers are the quarterback, trying to shape the action on the field based on constantly changing dynamics. The contracting officers are the center, no play starts until they hike the ball. They also defend the quarterback from getting crushed by oncoming contract protests, which are defensive tackles in this context.

❖ ❖ ❖

## The Government Art of Science

Another flavor of DOD customers are the Federally Funded Research and Development Centers (FFRDCs). They are unique public-private partnerships, designed to tackle specialized, long-term research and development needs for government agencies that don't always have a private market purpose. Born during World War II and refined in the Cold War era, FFRDCs emerged to harness scientific and technical talent outside the constraints of government bureaucracy (the irony of creating a government-funded entity to avoid bureaucracy is a recurring trope). They operate as independent, nonprofit entities often managed by universities or consortia. They are officially supposed to focus on emerging technologies that private industry does not have the incentive for basic and applied research. The federal government funds them, currently supporting 42 FFRDCs, with billions annually for R&D. Their mandate from Federal Acquisition Regulation § 35.017, emphasizes objectivity, access to sensitive data, and long-term stability, distinguishing them from profit-driven companies by barring them from competing for commercial work. This is an outdated belief, as FFRDCs absolutely compete with technology companies and have a built-in bias justified by an argument that the USG needs to have its own tech workforce in-house.

FFRDCs do have actual experts in several key defense fields, offering deep expertise and continuity. Their biggest, early developments were in mission-critical problems like radar advancements or nuclear weapons development from their WWII roots. Today, they span areas like aviation, space, and health, with sponsoring agencies (Department of Defense, Department of Energy, etc.) relying on them for strategic insights and technical innovation. Their structure is intended to provide freedom from shareholder pressures or political sway, which is arguably an impossible feat for organizations that must constantly sing for their supper funding.

An example is the RAND Corporation, the first FFRDC, established in 1947 by the U.S. Air Force. RAND focuses on policy analysis and decision-making support, tackling issues from military strategy to cybersecurity. Its independence should allow it to provide unbiased assessments, like evaluating defense systems or forecasting geopolitical risks, without the influence of corporate agendas.

These FFRDCs face tension with the high-tech industry, particularly over perceived competition. Critics, like the Professional Services Council, argue that private firms can match or exceed FFRDCs in areas like systems engineering and integration. Tech companies bristle at FFRDCs' privileged access to government data and sole-source contracts, claiming it creates an uneven playing field. For instance, while FFRDCs like MITRE advise on defense tech, private giants like Boeing or startups in Silicon Valley argue they could deliver faster, market-driven innovations if given the same shot. FFRDCs counter that their non-competitive status ensures impartiality and that private firms might prioritize profit over public interest, skewing outcomes. Congress has stepped in, as seen in the 2016 National Defense Authorization Act, pushing the DOD to scrutinize FFRDC roles and mitigate any unfair advantages. The clash hinges on a core question: can the private sector's agility and scale truly replace the FFRDCs' niche, trusted role in technology research.

❖ ❖ ❖

## Customers With Platinum Credit Cards

The "cool guys" like Navy SEALS, Rangers, and Green Berets have embraced their unique status and seem to have found ways to bend the acquisition system to their special desirements (this is a niche word in the industry reflecting a combination of desires and so-called "requirements".). US Special Operations Command and their more famous subordinate organization Joint Special Operations Command (pronounced J-Sock), play a pivotal role in shaping military innovation by writing requirements and procuring technologies that often ripple across the broader U.S. military. The many different Special Operations entities are famously accused of having platinum credit cards, allowing significant budget spend without the same level of oversight. JSOC was established in 1980 after the failed Iran hostage rescue mission exposed coordination gaps and was granted unique authorities under Title 10 to act as a quasi-service, managing its own budget and acquisition processes. This autonomy, paired with SOF's need for rapid, adaptable solutions in high-stakes missions, positions them as a trailblazer in developing and fielding technologies that later benefit conventional forces. From GPS integration during the Gulf War to today's advancements in ISR (intelligence, surveillance, and reconnaissance) and unmanned systems, SOF's influence is both practical and strategic, driven by their operational tempo and elite status.

SOF's role begins with defining precise requirements tailored to their unconventional missions—counterterrorism, direct action, special reconnaissance, and more. Unlike conventional forces, which often rely on large-scale, standardized systems, SOF has the advantage of experimenting with small quantities of custom equipment. This starts with operators and planners identifying gaps in capability during real-world deployments. For example, in the 1990s, SOF recognized the need for better navigation in featureless deserts, leading to the adoption and refinement of GPS technology. These requirements aren't drafted in a vacuum—SOF collaborates with engineers, scientists, and industry partners to translate battlefield needs into technical specifications. The process is iterative and fast-paced, often bypassing the Pentagon's cumbersome acquisition bureaucracy. USSOCOM's Special Operations Forces Acquisition, Technology, and Logistics (SOF AT&L) directorate oversees this, leveraging a $2 billion-plus annual procurement budget to prioritize speed over red tape. There is also plenty of overlap and budget alignment with the intelligence community, with varying amounts of "black budgets" that are not reported publicly.

Procurement is where SOF's unique authorities shine. Unlike the Army, Navy, or Air Force, which navigate layers of oversight, USSOCOM can directly contract with industry, often through rapid prototyping and small-batch production. This agility stems from its Major Force Program-11 funding, a dedicated line item that shields SOF from competing with service branches for resources. Their special aircraft, like the AC-130J and MH-60M show their ability to start with standard platforms and then completely outfit them for their unique missions. They sponsor an incubator "front door" organization called SOFWERX that tries to bring in small and non-traditional companies. Their short shelf-life of capabilities contrasts sharply with the conventional military's multi-year timelines, where years are spent just on the technical user manuals for hardware. SOF's smaller scale and mature operator base (averaging older, more experienced personnel) allow them to test and refine systems quickly, often in combat zones, before scaling up.

Once fielded, these technologies often transition to the wider military, a phenomenon driven by SOF's role as a "pathfinder." During the 1991 Gulf War, SOF's use of GPS to guide deep reconnaissance missions proved its worth. More recently, advancements in next-generation ISR, fusing cyber, space, and autonomous drones, emerged from SOF's need to track elusive targets around the world. USSOCOM's 2023 testimony to Congress highlighted investments in open-architecture command systems and space-based payloads, now eyed by the Space Force and broader DOD. The AC-130J gunship, recapitalized for SOF, is another example, with its precision firepower influencing Air Force upgrades. This handoff isn't accidental, SOF's testing in austere conditions provides a proof-of-concept, de-risking adoption for conventional units.

This process isn't without friction. SOF's focus on niche, high-cost solutions can spark tension with budget-conscious services prioritizing mass production. Critics argue SOF's bespoke approach diverts funds from joint priorities, as seen in debates over the Armed Overwatch program, a light attack plane tailored for permissive environments that some in Congress deemed redundant.

Historically, SOF's tech innovations reflect their operational DNA, small elite teams needing outsized impact. The Vietnam-era Air Force Pave Low helicopters, modified for night infiltration, set a precedent for today's stealth platforms. Post-9/11, SOF drove drone proliferation, with systems like the MQ-1C Gray Eagle now ubiquitous. Looking ahead, USSOCOM's focus on AI, cyber, and space, evident in partnerships with the Space Development Agency, positions SOF to shape future warfare.

# CHAPTER 3

# Defense Acquisition Lifecycle

ACQUIRING, MAINTAINING, AND EVENTUALLY REPLACING a large platform is not a single event or single contract. The worst visual PowerPoint slide ever created is Defense Acquisition University's "horse blanket" map of the entire process. The U.S. military's Joint Capabilities Integration and Development System (JCIDS) is a structured process to identify, assess, and prioritize warfighting needs, ensuring new capabilities align with strategic goals. Established in 2003 to replace service-specific requirements systems, JCIDS aims to foster jointness across the Department. Managed by the Joint Staff's J-8 Directorate, JCIDS interacts closely with the Planning, Programming, Budgeting, and Execution (PPBE) process, the DOD's resource allocation framework. Together, they shepherd multi-billion-dollar weapon systems, like the F-35 fighter or Virginia-class submarine, from concept to fielding. This directed process is so involved and tries to plan for every possible exception and step. Arguably, the defense acquisition system spends the majority of its time "checking boxes" and certifying compliance, with minimal investment in the actual purpose of the program.

JCIDS begins with a Capabilities-Based Assessment (CBA), where combatant commands, services, or agencies pinpoint gaps in doctrine, organization, training, materiel, leadership, personnel, facilities, or policy (DOTMLPF-P). For instance, a CBA might highlight inadequate missile defense against hypersonic threats. Analysts then draft documents like the Initial Capabilities Document (ICD), outlining the need, followed by a Capability Development Document (CDD) with detailed performance specs. These flow through the Joint Requirements Oversight Council (JROC), chaired by the Vice Chairman of the Joint Chiefs, which validates requirements for major systems (Acquisition Category I, or ACAT I, programs exceeding $524 million in R&D or $3.2 billion in procurement). The process emphasizes joint utility, ensuring a new tank or ship serves multiple services while balancing trade-offs in cost, schedule, and performance.

JCIDS feeds directly into PPBE, a four-phase cycle syncing requirements with funding. In the Planning phase, led by the Office of the Secretary of Defense (OSD), strategic guidance (e.g., the National Defense Strategy) shapes priorities, informed by JCIDS outputs. Programming, managed by service components and OSD's Cost Assessment and Program Evaluation (CAPE) office, translates validated requirements into multi-year Program Objective Memorandums (POMs), detailing resource needs like $10 billion for a new destroyer class over five years. Budgeting refines these into annual budget submissions, scrutinized by OSD and sent to Congress. Execution, the final phase, allocates funds and tracks spending as programs unfold. PPBE's calendar-driven rhythm forces JCIDS to align its often-slower requirements timeline, creating friction when urgent needs arise mid-cycle.

For exquisite systems like the $100 billion B-21 Raider program, JCIDS and PPBE face daunting schedule and budget hurdles. First, requirements creep plagues JCIDS. Stakeholders pile on "nice-to-have" features, ballooning complexity. The F-35's CDD, for example, demanded stealth, vertical takeoff, and multi-role capabilities across three variants, driving costs from $59 million per jet in 1994 estimates to over $100 million today. JROC's joint focus can exacerbate this, as services negotiate to protect turf, delaying consensus. Second, schedule misalignment between JCIDS and PPBE stalls progress. JCIDS might take 18-24 months to validate a need, but PPBE's budget lock-in occurs years ahead, forcing programs to wait or rush with incomplete specs.

Budget challenges compound these issues. PPBE's fixed topline ($859 billion for Fiscal Year 2025) pits exquisite systems against other priorities like personnel or readiness. CAPE's cost estimates, meant to ground JCIDS specs in reality, are like all other estimates relying on assumptions. The Navy's Gerald R. Ford carrier, initially pegged at $10.5 billion, hit $13.3 billion by 2017 due to untested tech like electromagnetic catapults, exposing PPBE's struggle to fund "concurrency" building while designing. Congress adds pressure, cutting funds (e.g., trimming F-35 buys in FY 2023) or mandating reviews, as with the 2016 NDAA's scrutiny of cost overruns. Meanwhile, exquisite systems face technological risk. Pushing boundaries, like hypersonic missiles or directed-energy weapons, means betting on unproven science, inflating timelines from 10 to 20 years.

External factors amplify these woes. Adversaries like China iterate faster, fielding systems in half the time, while the

U.S. processes plods along. GAO reports slam JCIDS/ PPBE for delivering late, over-budget platforms. F-35s arrived a decade behind schedule, still lacking full capability. Industry chafes too: Lockheed Martin or Northrop Grumman must front R&D, only to see requirements shift or budgets shrink, squeezing margins. Reform efforts, like the 2022 Adaptive Acquisition Framework, push modular, open-system designs to speed delivery but cultural inertia, services clinging to bespoke solutions, slow change.

Case studies highlight the stakes. The Virginia-class submarine streamlined requirements for a post-Cold War attack sub, hitting budget ($2.8 billion per boat) and schedule (first delivered 2004) via disciplined trade-offs. Contrast that with the Littoral Combat Ship (LCS), a $37 billion fiasco. JCIDS approved a flexible, shallow-water vessel, but vague specs and PPBE's rushed funding led to two incompatible designs, maintenance woes, and early retirement. The difference? Clear requirements and fiscal realism versus ambition outpacing execution.

Ultimately, JCIDS and PPBE aim to balance capability with needs, but exquisite systems test their limits. Schedule delays erode deterrence, GAO notes 85% of ACAT I programs miss milestones, while budget overruns drain resources. The DOD has spent $2 trillion on acquisitions since 2000, according to the Congressional Budget Office. Fixes like agile acquisition or prototyping (e.g., SOF's rapid cycles) offer hope, but scaling them to billion-dollar behemoths remains elusive. As threats evolve, the DOD must wrestle with whether exquisite perfection is worth the cost or if simpler, faster solutions better serve the fight.

In April 2025 the Trump Administration surprised most of the industry and defense enterprise with an ambitious Executive Order directing a reform of the entire DOD acquisition process strengthening the preference for commercial offerings and the use of more flexible Other Transaction Authority contract types.

❖ ❖ ❖

## Phases of the Defense Acquisition Lifecycle

The U.S. Department of Defense Acquisition Lifecycle unfolds as a carefully orchestrated journey (or a slow-moving, repetitive car crash, depending on your view), brought to life by Program Executive Officers and Program Managers who work hand-in-hand with the sprawling defense industry. This process, guided by DOD Instruction 5000.02 and the Adaptive Acquisition Framework, moves through distinct chapters: Material Solution Analysis, Technology Maturation and Risk Reduction, Engineering and Manufacturing Development, Production and Deployment, and Operations and Support. Each chapter ends with a Milestone review, A through C, where a Milestone Decision Authority gives the nod to proceed.

PEOs and PMs serve as the story's protagonists, steering the creation of military systems from raw need to battle-field reality, while industry giants like Lockheed Martin and Northrop Grumman provide the muscle and ingenuity to make it happen. The Program Managers chart the course, aligning acquisition with the service's grand vision, securing funds through PPBE and Congressional Appropriations.

In a "traditional" program, it begins with a Material Solution Analysis, where PEOs and PMs confront a gap flagged by JCIDS. The PM rallies a team, leaning on industry voices like Raytheon to weigh options: upgrade old systems or forge something new. At Milestone A, the PEO locks in the strategy and initial funding. The plot thickens in Technology Maturation and Risk Reduction, as PMs collaborate with firms like Boeing to prototype and tame risks, maybe testing a jet's stealth coating. The PEO shepherds this effort toward Milestone B. Engineering and Manufacturing Development follows, with PMs and contractors refining designs and running trials, while PEOs keep budgets and timelines in check. Milestone C ushers in Production and Deployment, where PMs oversee factories humming with activity. The final act, Operations and Support, sees PMs sustain systems through industry logistics, like Huntington Ingalls, constantly maintaining and updating ships.

The defense industry is not a casual observer to the lifecycle. In Material Solution Analysis, companies like BAE Systems pitch studies or ready-made fixes. Technology Maturation and Risk Reduction makes choices on which "good ideas" from stakeholders and industry the program can afford. Engineering and Manufacturing Development hands a lead systems integrator the reins to build and troubleshoot proto-types. Production and Deployment shift to mass output at either a company site like BAE's ground vehicle center in York, Pennsylvania or a Government-Owned facility like the Joint Systems Manufacturing Center in Lima, Ohio. Industry doesn't just follow orders, it shapes the script.

Challenges are constant for PEOs and PMs. Schedules stretch long with changes, GAO noting 85% of big programs lag, as industry grapples with unproven tech in Technology Maturation and Risk Reduction. PEOs wrestle with PPBE's stiff deadlines, often committing funds before designs gel. Budgets balloon when industry underbids, LCS's $220 million fantasy morphing into a $37 billion beast, leaving PEOs begging Congress and PMs haggling fixes.

❖ ❖ ❖

## The Program Objective Memorandum (POM) Cycle

The POM cycle begins with the Defense Planning Guidance, issued by the Office of the Secretary of Defense. This document outlines strategic goals, threat assessments, and budgetary limits, reflecting priorities from the National Defense Strategy. Each military service, including the Army, Navy, Air Force, Marine Corps, and Space Force, along with defense agencies, develops its own POM. These submissions propose funding allocations for programs, personnel, training, equipment, and infrastructure, balancing immediate operational demands with long-term investments in capabilities like advanced weaponry or cybersecurity.

Running from approximately May to August, the POM cycle involves rigorous coordination and analysis. Services craft their POMs based on internal assessments of mission requirements and cost estimates, often using data from acquisition programs and operational plans. These drafts are submitted to the Office of the Secretary of Defense, where the Director of Cost Assessment and Program Evaluation leads a review. This phase identifies discrepancies, evaluates trade-offs, and ensures alignment with overarching defense goals. Analysts scrutinize program costs, schedules, and performance metrics to optimize resource distribution.

Stakeholder collaboration is central to the process. Program review teams, comprising representatives from the services, Joint Staff, and combatant commands, address contentious issues, such as competing funding priorities or program overlaps. Senior leaders, including the Deputy Secretary of Defense, resolve disputes through Program Decision Memoranda, which finalize adjustments to the POMs. These decisions shape the ~3,600 individual Budget Activity submissions, which feed into the broader budget formulation process for congressional review.

The POM cycle faces challenges, including adapting to rapidly evolving threats, managing budget uncertainties, and incorporating cutting-edge technologies like artificial intelligence. Congressional oversight and fiscal caps, such as those imposed by the Budget Control Act or Fiscal Responsibility Act, add complexity. Delays in passing annual appropriations can disrupt planning, forcing reliance on continuing resolutions that limit flexibility. The idea of Congressional appropriations being directed all the way deep into funding lines by program was not how the process started.

❖ ❖ ❖

## The Palette of Defense Spending: Understanding the Colors of Money

In U.S. defense appropriations, the concept of "Colors of Money" refers to the distinct categories of funding allocated by Congress, each with specific purposes, rules, and timelines. These categories ensure the Department spends taxpayer dollars as intended, tying funds to activities like research, procurement, or operations. The term "colors" is a shorthand for the legal and budgetary boundaries that PEOs and budget officials are forced to stay in line with. Misusing these funds risks audits, penalties, or program delays. Defining different types of spend with severely restrictive ability for re-programming perversely incentivizes rigid stagnation in programs. If a five percent increase in R&D could save 25% in long-term maintenance, there is no definite way to achieve that trade-off.

The main "colors" include Research, Development, Test, and Evaluation (RDT&E), Procurement, Operations and Maintenance (O&M), Military Personnel (MILPERS), and Military Construction (MILCON). Each has a unique lifespan and scope, set by Congress through annual bills, like the National Defense Authorization Act (NDAA) and accompanying funding bills. RDT&E, with a two-year lifespan, fuels innovation. It covers designing and testing systems, like hypersonic missile prototypes at Lockheed Martin, costing $41 billion across DOD in FY 2023. Procurement, valid for three years, buys hardware, like F-35 jets or Virginia-class submarines. O&M, a one-year fund, keeps the military running, paying for fuel, training, and base upkeep. MILPERS, also one-year, funds salaries and benefits for troops while MILCON, with a five-year window, builds facilities like barracks or runways.

These colors emerge from the PPBE cycle. During Programming, services craft their POM submissions, splitting needs into these buckets. The Navy might request $2 billion in Procurement for a destroyer, $500 million in RDT&E for its radar, and $300 million in O&M for crew training. Congress locks these into law, and the DOD's Financial Management Regulation (DOD 7000.14-R) enforces strict adherence to the codified plan. RDT&E cannot buy finished gear, Procurement cannot pay salaries, and O&M cannot fund construction. This rigidity favors transparency and oversight over actual buying power of capability per dollar spent.

PEOs and PMs wrestle with these constraints daily. A PM developing a drone might use RDT&E to test prototypes, then shift to Procurement for production, timing the switch to match Milestones in the Acquisition Lifecycle. If RDT&E runs dry before Milestone B, the program stalls unless Congress reprograms funds, a slow process needing approval. O&M, expiring yearly, forces rushed spending or lost funds, dubbed "use it or lose it." MILCON's long horizon suits multi-year builds, but delays, like a late environmental review, can tie up cash. The colors' lifespans, one to five years, reflect their purpose: short for operations, longer for capital investments.

The defense industry feels this, too. Contractors bid on RDT&E contracts for early-stage work, then build capacity for full-rate production during the first couple phases. Misaligned funding can often force a Service, and the vendor into making hasty choices to force the technology to align with the date the procurement budget shows up from Congress. Firms often front costs, betting on future appropriations, a gamble if Congress cuts back. Industry pushes for stable, multi-year funding, but annual budgets and political shifts disrupt plans.

Improving the ability for programs to shift funding is a major reform topic that comes up almost every year. A sudden escalation, like the Houthi attacks in the Middle East, might need Procurement funds for missiles, but O&M cannot shift without congressional action, a months-long slog. The Government Accountability Office flags frequent "color violations," like using O&M for minor buys meant for Procurement, triggering audits. The budget splits tightly across categories, sparking service rivalries, Navy ships versus Air Force planes. Reform efforts, like the 2022 PPBE Commission's call for flexibility, aim to ease this, but tradition holds firm.

CHAPTER 4

# The Big Players

## *Prime Contractors, Consolidation, and Everyone's 401k*

*"It was natural that we would become a bit complacent."*
—Thomas G. Pownall, President of
Martin Marietta Aerospace Group, May 18, 1971

MASSIVE "PRIME" CONTRACTORS ARE NOT a flaw in the U.S. defense system but a distinct feature created by a combination of economic and political choices. The sheer scale and complexity of major acquisitions have pushed the industry toward corporations capable of handling a vast array of tasks to meet the demands of an ever-expanding defense acquisition framework. These primes represent a distinctly American solution to a unique challenge. Their dominance intensified after the "Last Supper," a 1993 meeting where the Pentagon urged consolidation to streamline supposed Cold War-era excess, reshaping the industry into a handful of giants. This chapter explores the key prime contractors, tracing their origins and distinct traits, while addressing growing customer frustration that many have swelled into "too big to fail" behemoths.

The attack on Pearl Harbor is often credited with launching the "Arsenal of Democracy" as President Roosevelt described the shift from consumer production to defense materiel. *Freedom's Forge* by Arthur Herman explains how this shift by Ford, Chrysler, and others had already been converting to wartime production to support the lend-lease program. In the early 1940s, this powerhouse churned out 100,000 tanks, 300,000 aircraft, and 17 million firearms, a wartime surge that forged the foundation of today's defense industry, influencing politics, economics, and society for generations.

Today, that legacy endures in titans like Lockheed Martin, Boeing, and Raytheon. "Blue Chip" firms with enterprise values climbing into the hundreds of billions. They form the backbone of what President Dwight Eisenhower labeled the "military-industrial complex." Lockheed Martin, the pin-striped Yankees of the defense industry, posted $71 billion in revenue in 2024, outstripping the GDP of several states and even some small countries. Its F-35 Joint Strike Fighter program, with a lifetime cost exceeding $1.7 trillion, underscores its colossal reach.

These companies are very unique in that their products and services are arguably essential but their ability to earn revenue is almost completely dictated by 535 members of both chambers of Congress. They are woven into America's fabric, employing hundreds of thousands nationwide. In some congressional districts, they account for 20 percent of jobs, a testament to how the "Arsenal of Democracy" evolved from a wartime sprint into a peacetime pillar.

Their sway stretches beyond production lines and battle-fields. Millions of Americans hold their stocks in retirement accounts, with State Street Corp., Vanguard, and BlackRock owning over 81 million Lockheed Martin shares alone. This ties the average citizen, far removed from military bases, to the war business. When defense stocks climb, IRAs swell; when contracts land, micro-regions can boom, making most Americans silent stakeholders.

The SPADE Defense Index tracks publicly traded companies vital to U.S. defense, homeland security, and aerospace sectors. Launched in 2004 with data back to 1997, it serves as a benchmark for firms driving nearly 4 percent of U.S. GDP. The index uses a modified market capitalization weighting, capping large companies to balance influence, and includes players like Boeing, Lockheed Martin, and newer entrants like Palantir. It covers manufacturers of aircraft, drones, missiles, and cybersecurity systems, adapting to shifts in military needs, such as AI and hypersonics. To qualify, firms need a minimum $250 million market value, a $5 share price, and $10 million in quarterly revenue. The index rebalances quarterly, adding or dropping companies based on market trends, mergers, or performance. Traded as DXS on the NYSE, it powers the Invesco Aerospace & Defense ETF, offering investors exposure to a sector fueled by $2 trillion in global defense spending. Its long-term performance has outpaced broader markets, reflecting steady growth despite budget fluctuations.

The concept of defense primes started with World War II production shifts, when Chrysler built 25,000 Sherman tanks and General Motors' Willow Run plant rolled out a B-24 bomber hourly. Victory cemented their role, but peacetime and the Cold War redefined it. The 1950 Defense Production Act, sparked by the Korean War, empowered the government to steer industry for defense, a tool still used today for crises like ventilator shortages in 2020.

The primes vary in roots and style. Lockheed Martin, born from a 1995 merger of Lockheed and Martin Marietta, thrives on jet aviation like the Joint Strike Fighter. Boeing, a commercial aviation pioneer, balances defense with passenger jets. In early 2025 they were announced as the winner to build a new sixth-generation Air Force platform, reinvigorating their defense portfolio. Raytheon, now RTX, excels in missiles and electronics, its Patriot system is a global staple.

These primes, deemed "too big to fail," wield outsized influence. There is a constant debate over whether the requirements or budget provided by the DOD was ever aligned with reality, or whether the prime contractor over-promised and under-delivered. Consolidation cuts competition, stunting innovation, while their economic clout pressures lawmakers to sustain them. As cyber, space, and AI reshape warfare, new players like SpaceX and Palantir challenge the old guard, hinting at a shift. The next "Arsenal" will likely come from Silicon Valley's software engineers.

❖ ❖ ❖

## The Defense Industry's Last Supper: A Watershed Moment

The "Last Supper" marked a pivotal moment in the U.S. defense industry's evolution, setting off a wave of consolidation that reshaped the sector and sparked decades of debate. Held in 1993 at the Pentagon, this dinner brought together Defense Secretary Les Aspin and top executives from major defense contractors. The Cold War had ended, the Soviet Union had collapsed, and military spending faced steep cuts. Aspin delivered a blunt message: the industry had too many players to survive the lean years ahead. He urged companies to merge or risk extinction, arguing that consolidation would streamline costs and keep the industrial base viable. What followed was a frenzy of mergers and acquisitions, shrinking a once-diverse field of over 50 firms into a handful of giants, a process that unfolded over years and raised questions about competition, innovation, and national security.

Before the "Last Supper," the defense industry thrived on Cold War budgets, supporting dozens of companies building everything from tanks to missiles. Aspin, a former congressman with a keen eye for defense policy, saw this sprawl as unsustainable. His logic was simple: fewer, larger firms could weather the downturn, pooling resources to maintain capability. Industry leaders took the hint. Within a decade, mergers fused names like Lockheed and Martin Marietta into Lockheed Martin, Northrop and Grumman into Northrop Grumman, and Boeing absorbed McDonnell Douglas. By 2000, the top five primes controlled over 70 percent of DOD contracts, a stark shift from the fragmented market of the 1980s.

This consolidation wasn't just a business shuffle; it altered the industry's DNA. Lockheed Martin, reborn in 1995, became a titan. Boeing, already a commercial giant, bolstered its defense arm with McDonnell Douglas in 1997, balancing jets and bombers. Northrop Grumman, merging in 1994, leaned into stealth and electronics, later expanding through further acquisitions like Orbital ATK. The "Last Supper" didn't mandate these moves, but Aspin's nudge, backed by FTC tolerance of mergers, ignited them. Critics argue it traded true competition for lethargic reliance on risk avoidance and compliance with checklists.

A controversial chapter in this saga came with Northrop Grumman's 2018 acquisition of Orbital ATK, a deal the Federal Trade Commission (FTC) approved despite red flags. Orbital ATK, formed in 2015 from Orbital Sciences and Alliant Techsystems, was one of two major U.S. solid rocket motor suppliers, alongside Aerojet Rocketdyne. Its $9.2 billion buyout by Northrop Grumman raised eyebrows, as it left the rocket motor market with just two players, a near-duopoly critics feared would choke competition. The FTC greenlit it with conditions: Northrop had to maintain Orbital as a "merchant supplier," selling motors to rivals like Lockheed Martin without favoritism. Skeptics saw this as flimsy. During the $80 billion Ground Based Strategic Deterrent competition, Northrop firewalled Orbital to level the playing field, but competitors dropped out, leaving GBSD a sole-source contract in 2020. Costs climbed, innovation lagged, and whispers of cozy Pentagon-industry ties grew. The approval, under a Trump-era FTC less aggressive on antitrust, fueled suspicions that national security pleas, like missile stockpile needs, swayed regulators over market health.

Fast forward to 2023, and the L3Harris acquisition of Aerojet Rocketdyne stirred fresh debate, echoing the Orbital ATK fallout. Aerojet Rocketdyne, the last independent rocket motor maker after Orbital's absorption, had struggled since the FTC blocked Lockheed Martin's $4.4 billion bid in 2022. That rejection, under Biden's tougher antitrust stance, cited vertical integration risks: Lockheed, a missile prime, owning Aerojet, could squeeze rivals' access to motors. L3Harris, a $17 billion firm formed in 2019 from L3 Technologies and Harris Corp, swooped in with a $4.7 billion offer in December 2022. Unlike Lockheed, L3Harris wasn't a direct missile competitor, framing the deal as "horizontal," adding propulsion to its portfolio without overlapping markets.

The FTC scrutinized this too, reflecting post-"Last Supper" wariness of consolidation. Aerojet's $2.3 billion revenue and role in hypersonics, missiles, and space made it a prize, but its duopoly status with Northrop Grumman still caused consolidation fears. Senator Elizabeth Warren urged rejection in January 2023, warning it would "reduce competition to a new low," hiking prices and stunting innovation. L3Harris countered that it would stay a neutral supplier, selling motors to all primes, not locking them up like Lockheed might have. The FTC requested more data in March 2023, extending the review, but by July, it opted not to block the deal. L3Harris closed the acquisition on July 28, 2023, absorbing Aerojet as a fourth business unit.

Supporters saw it stabilizing Aerojet, reeling from boardroom fights and delivery struggles. The Pentagon, eyeing Ukraine-driven motor demand, awarded Aerojet $216 million in April 2023 to expand plants, hinting at tacit approval. Critics, though, saw echoes of Orbital ATK: another step toward monopoly, with L3Harris now wielding leverage in a thin market.

The "Last Supper" launched this arc, shrinking a vibrant industry into a club of giants. Orbital ATK's acquisition showed how consolidation could distort competition, its FTC pass raising doubts about oversight. L3Harris's Aerojet grab, while cleared, deepened unease that Aspin's vision birthed a system too concentrated to pivot fast. Today's primes, employing hundreds of thousands and tied to 401(k)s via 81 million Lockheed shares held by funds like Vanguard, are economic linchpins. As cyber and space threats loom, voices like Christian Brose's in his book *Kill Chain* question if these giants, wedded to legacy systems, can match agile rivals. With Christian Brose now serving as the President and Chief Strategy Officer of Anduril, it shows that there are new primes challenging the status quo.

❖ ❖ ❖

## Bombers and Fighter Jocks

Boeing, Lockheed Martin, and Northrop Grumman stand as pillars of the U.S. defense industry, each with distinct origins, iconic programs, and M&A histories that shaped their dominance. Born from different eras and ambitions, these primes evolved through innovation, wartime necessity, and strategic consolidation. Their famous programs, from bombers to stealth fighters, showcase their technical legacies, while their M&A paths reveal how they navigated shrinking budgets and rising competition. Comparing these giants highlights a blend of shared reliance on defense contracts and unique trajectories tied to their founding visions.

Boeing's story began in 1916, when William Boeing, a timber magnate turned aviation enthusiast, founded the Pacific Aero Products Company in Seattle. Renamed Boeing Airplane Company in 1917, it started with seaplanes for the U.S. Navy during World War I. The company pivoted to commercial aviation in the 1930s, launching the 247 airliner, but World War II cemented its defense roots. Boeing's B-17 Flying Fortress and B-29 Superfortress bombers, with over 12,000 and 3,900 built respectively, became icons of Allied air power. Post-war, it balanced civilian and military work, rolling out the 707 jetliner in 1958 and the B-52 Stratofortress (still flying today and likely to have a 100-year service life). The Apollo program's Saturn V first stage further showcased their versatility. Famous programs like the 737, with over 10,000 sold commercially, and the KC-46 Pegasus tanker, a $44 billion Air Force deal, highlight Boeing's dual-market strength. Now in 2025 they have been awarded the Next Generation Air Dominance platform by the Air Force, strangely numbered as the F-47.

Lockheed Martin traces its lineage to 1912, when brothers Allan and Malcolm Loughead formed the Alco Hydro-Aeroplane Company in California, later renamed Lockheed Aircraft Company in 1926. Early success came with the Vega, a 1920s airliner flown by Amelia Earhart, but World War II propelled it forward. The P-38 Lightning fighter, with 10,000 produced, shone in the Pacific, while the Cold War birthed legends like the U-2 spy plane and SR-71 Blackbird (still the fastest manned aircraft at Mach 3.2). The 1995 merger with Martin Marietta, a 1961 union of Martin Company and American-Marietta, created Lockheed Martin, blending Lockheed's aviation prowess with Martin's missile expertise. Its crown jewel, the F-35 Joint Strike

Fighter, a $1.7 trillion program with over 1,000 delivered by 2025, has been a staple of NATO aviation.

Northrop Grumman's roots lie in 1939, when Jack Northrop, a visionary designer, founded Northrop Corporation in Hawthorne, California. Building on his 1920s work with Lockheed, Northrop chased radical designs like the YB-35 flying wing bomber, a 1940s flop reborn as the B-2 Spirit stealth bomber. World War II saw it produce P-61 Black Widow night fighters, with 706 built, but its fame grew later. The 1994 merger with Grumman Corporation, founded in 1929 by Leroy Grumman and known for Navy staples like the F6F Hellcat (12,275 units) and F-14 Tomcat, formed Northrop Grumman. The B-2, with 21 units at $2 billion each, and the B-21 Raider, a $203 billion next-gen bomber program in development, define its stealth niche. Its space and missile defense work, like the James Webb Space Telescope, adds depth.

Boeing's M&A history reflects opportunistic growth. The 1997 $13.3 billion merger with McDonnell Douglas, a 1967 fusion of McDonnell Aircraft and Douglas Aircraft, bolstered its defense portfolio with F-15 Eagles and Delta rockets, countering post-Cold War cuts. Earlier, the 1960 Vertol acquisition added helicopters like the CH-47 Chinook, with over 1,200 built. Recent moves, like the 2018 Embraer joint venture attempt (scrapped in 2020), aimed at commercial gains, showing Boeing's focus on balancing sectors. It's $4 billion in 2023 defense M&A targeted tech like Aurora Flight Sciences for autonomy.

Lockheed Martin's M&A path, turbocharged by the "Last Supper," centers on scale. The 1995 Lockheed-Martin

Marietta deal, a $10 billion blockbuster, merged airframes with missiles, absorbing Martin's Titan and Viking programs. The 1996 $9.1 billion Loral Corporation buy added electronics, enhancing C4ISR capabilities. Later, the 2015 $9 billion Sikorsky Aircraft purchase from United Technologies brought helicopters like the UH-60 Black Hawk, with over 4,000 built, into its fold. For a deep analysis of Lockheed Martin there are two seminal books that go over the history and unique engineering prowess.

*Skunk Works*, written by Ben R. Rich and Leo Janos, chronicles the history of Lockheed Martin's secretive Skunk Works division, renowned for designing revolutionary aircraft. Rich, a former Skunk Works director, details the division's innovative culture, emphasizing rapid prototyping, minimal bureaucracy, and close-knit teams under founder Kelly Johnson. The book recounts engineering triumphs, such as the SR-71's unmatched speed and the F-117's stealth technology, alongside challenges like tight budgets, intense secrecy, and Cold War pressures. It highlights Johnson's 14 rules for efficient project management, which prioritize simplicity and autonomy. Rich also shares personal anecdotes, illustrating the human side of high-stakes defense projects. The narrative underscores Skunk Works' role in advancing U.S. military capabilities and its lasting influence on aerospace innovation. Published in 1994, the book remains an influential account of defense industry ingenuity and covert engineering excellence.

At the other end of the spectrum, providing a largely critical analysis of Lockheed Martin is *Prophets of War* by William D. Hartung. He raises questions about what kind of force a behemoth like Lockheed Martin exerts on US foreign policy and whether the outsourcing of so much warfare capacity to a publicly traded company can ever remain honest.

A 1994 $2.1 billion Grumman acquisition fused Northrop's stealth with Grumman's naval legacy, like the E-2 Hawkeye. The 2001 $5.2 billion Litton Industries and Newport News Shipbuilding buy added shipyards. The 2018 $9.2 billion Orbital ATK acquisition, blending Orbital's satellites and ATK's rocket motors, stirred FTC scrutiny but passed with conditions, securing Northrop's GBSD sole-source win. Its $78 billion market cap underscores this growth.

Famous programs highlight these identities. Boeing's B-52 and KC-46 blend longevity and utility, serving broad Air Force needs, while its 737 ties it to airlines. Lockheed Martin's F-35 and SR-71 push tech frontiers, feeding its high-stakes reputation. Northrop Grumman's B-2 and B-21, plus space feats like Webb, carve a futuristic niche. Boeing's volume, Lockheed Martin's ambition, and Northrop Grumman's stealth define their legacies.

M&A strategies diverge too. Boeing's measured grabs, like McDonnell Douglas, bolster both wings, though recent stumbles (737 MAX) shift focus. Lockheed Martin's post-1993 spree, from Loral to Sikorsky, chases dominance, hitting FTC walls with Aerojet. Northrop Grumman's targeted buys, like Orbital ATK, deepen specialties, balancing growth with agility. Boeing spreads wide, Lockheed Martin stacks tall, Northrop Grumman digs deep.

Common threads bind them: all leaned on World War II to scale, rode Cold War budgets, and consolidated after Aspin's nudge. Boeing's commercial anchor, Lockheed Martin's defense purity, and Northrop Grumman's tech focus set them apart.

❖ ❖ ❖

## Changing of the Guard

The defense industry has long been dominated by the defense primes, which have secured the majority of the Department of Defense's RDTE and procurement budgets These legacy contractors have built deep relationships with the Pentagon, leveraging their mastery of complex regulations and procurement processes to maintain an oligopoly. There is a growing shift in preference toward new disruptors like Palantir and Anduril, which are challenging the traditional model with innovative, software-driven solutions. This shift is compounded by external pressures, including the Trump administration's constantly changing executive orders. The primes have long relied on the implied "warranty" of US involvement for any global purchasers of their hardware.

Historically, prime contractors have excelled in delivering complex, hardware-intensive systems that take years to design and produce. Lockheed Martin's F-35 program is the world's largest defense project, with a lifetime cost estimated at $1.7 trillion. These firms benefit from entrenched relationships with the Pentagon, extensive lobbying networks, and contracts structured to favor their capabilities. Their dominance has been criticized for fostering inefficiencies, high costs, and slow adoption of new technologies, as the procurement process often prioritizes established players over innovative newcomers. This has created a barrier for smaller firms trying to cross the "Valley of Death" from nascent technology to a full program of record.

Enter disruptors like Palantir and Anduril, which represent a new breed of defense contractors focused on software, artificial intelligence, and autonomous systems. Palantir, co-founded by Peter Thiel, specializes in data analytics and AI platforms, such as its work on the Army's TITAN targeting system, which integrates sensor data for rapid battlefield decision-making. Anduril, cofounded by Palmer Luckey, develops autonomous drones and counter-drone systems, emphasizing agility and cost-effectiveness. These companies have gained traction by addressing modern military needs, such as real-time data processing and uncrewed systems, which are increasingly vital in conflicts like those in Ukraine and the Middle East. Their commercial tech roots enable faster development cycles and potentially lower costs compared to traditional contractors, appealing to a Pentagon seeking to modernize.

Palantir, Anduril, and SpaceX are actively challenging the status quo by teaming to chase the recently announced "Golden Dome for America." Palantir's massive growth in 20 years signals a strategic push to reallocate defense budgets toward smaller, tech-savvy firms. Palantir's market capitalization, surpassing Lockheed Martin's at $169 billion, and SpaceX's $350 billion valuation underscore the growing economic clout of these disruptors.

This shift is amplified by the Trump administration's policies, which seem to be willing to upset any apple cart with a preference for taking immediate action with broad, sweeping executive order pronouncements before the DOD has the opportunity to draft implementing instructions. Key figures like Vice President JD Vance, a former associate of Peter Thiel, and Elon Musk advocate for redirecting Pentagon funds to innovative startups. Musk's criticism of legacy programs, particularly the F-35, aligns with calls for cheaper, AI-powered alternatives like drones and autonomous submarines. Defense Secretary Pete Hegseth has also criticized the Pentagon's insularity, praising Silicon Valley's contributions. These dynamics have fueled optimism among defense tech startups, while traditional contractors like Lockheed Martin have seen share prices slump post-election.

However, the Trump administration's broader policies are creating significant hurdles for legacy contractors. Threats to withdraw from NATO unless member states increase defense spending have strained alliances, significantly reducing demand for U.S. defense exports. Unprecedented changes in tariff and trade policy will have seismic effects on the industry. Lockheed Martin, which relies heavily on international sales for programs like the F-35, faces risks if NATO countries scale back purchases due to political tensions or budget constraints.

Tariffs and trade embargo responses from other nations on critical materials such as steel, aluminum, copper, and rare earth minerals threaten to disrupt supply chains, mirroring COVID-era shortages. These tariffs, intended to bolster domestic production, will raise costs and delay programs. Analysts warn that such disruptions could exacerbate existing challenges, including spare parts shortages and aging aircraft fleets.

Despite these pressures, traditional contractors retain advantages in navigating Pentagon bureaucracy and delivering large-scale systems. While disruptors excel in software and autonomy, they lack the infrastructure to produce complex hardware like stealth fighters or naval vessels. This physical capacity gap will not last long, Anduril is investing $1 billion in their own "Arsenal-1" as a "hyperscale manufacturing facility." The Pentagon's continued investment in conventional systems suggests that legacy firms will not be displaced entirely. However, the growing influence of tech-driven firms, backed by political allies in the Trump administration, indicates a transformative period for the defense industry. For Lockheed Martin and its peers, adapting to this new landscape while managing tariff-induced costs and geopolitical uncertainties will be critical to maintaining their relevance.

# CHAPTER 5

# Everyone Else

## Main Street, Small Business, and Silicon Valley

*"...not only the wealth but the independence and security of a country appear to be materially connected to the prosperity of manufacturers."*
—Alexander Hamilton,
Report on Manufactures, 1791

BEYOND THE QUARTERLY EARNINGS REPORTS and headlines about multi-billion-dollar platform contract awards there is an entirely different ecosystem of players in the defense industry. From "Main Street" businesses crafting niche components to venture capital firms betting on disruptive tech, private equity groups reshaping mid-tier suppliers, and international companies that dabble around the edges of the defense industry enterprise. All of the players in the defense sector are not created equal, they are not all trying to compete with Lockheed and Raytheon, and there is a lot of smart analysis showing that the supply chain modernization to just-in-time and international dependencies has eroded our supply base.

"Main Street" businesses, the small and medium-sized companies dotting towns from Ohio to Oregon supply fasteners, bolts, light bulbs, welded fixtures, and literally millions of other items and components that feed up the chain. These aren't flashy primes but machine shops, electronics makers, and family-owned outfits, often with fewer than 50 employees. They churn out precision parts like gears, rods, and assemblies. The Defense Logistics Agency, which serves as the Amazon Fulfillment Center of spare and repair parts for the DOD, does the most contracting with these small main street businesses. Firms like Whelan Machine and Tool in Louisville thrive on subcontracts from primes and hard-to-find direct DOD opportunities. These small companies wrestle with razor-thin margins, bureaucratic hurdles like CMMC data security compliance, dependence on prime whims, and now major uncertainty with tariffs and war on supposed "government waste."

Venture capital injects a different energy, making major swings into defense technology (shortened to DefTech). Silicon Valley's Anduril, founded in 2017 by Palmer Luckey with VC backing from Founders Fund, exemplifies this shift. Its Lattice AI platform and Ghost drones attempt to bypass traditional acquisition sluggishness, landing $1.5 billion in contracts by 2024. VC firms like Bessemer Venture Partners and Lux Capital are investing billions into startups in defense, or at least with a strong case in the national security sector. Unlike primes wedded to legacy systems, these startups pitch speed and scalability, often sidestepping JCIDS for rapid prototyping via flexible Other Transaction Authority contracts. The VC challenge in defense is crossing the infamous "valley of death."

Private equity carves its own niche, reshaping the industry's middle tier. Firms like Carlyle Group, KKR, and Arlington Capital Partners scoop up suppliers and service providers, streamlining them for economy of scale, then selling or merging them into bigger players. Arlington Capital's roll-up of over 16 companies under the name BlueHalo ended with a major sale to AeroVironment in 2024. PE invested over $5.6 billion in defense in 2023 targeting firms like Triumph Group, a Boeing supplier PE rescued from bankruptcy in 2020. These deals inject capital and efficiency, often modernizing dusty operations, which primes overlook. Critics argue PE's short-term focus can gut R&D or saddle firms with debt.

International companies add a global flavor, bridging U.S. needs with foreign expertise. Britain's BAE Systems rolled up several US companies including the famous combat vehicle producer United Defense. With over $26 billion in 2024 revenue, they are one of the largest US primes. Norway's Kongsberg produces Naval Strike Missiles and remote weapon stations. The Foreign Military Sales (FMS) program has been a foundation of US international relations policy, trading capability and hardware for economic and co-defensive alignment.

❖ ❖ ❖

## Defense Industry Contributors

There are many large companies that are absolutely part of the broader definition of Defense Industrial Base that may not consider themselves "defense" companies. Companies like 3M, Dow, and Dupont have all contributed materials and products. These companies have existed across eras when they were default suppliers to the defense apparatus, but then suffered as the public market appetite for conglomerates contributing chemicals and components to the war industry became liabilities. Arguably, the professionalization of the acquisition and procurement field in the DOD started to favor "specialist" defense-pure players in the 1970s.

Throughout the 1980s, large U.S. companies, particularly those in aerospace, technology, and manufacturing, strategically reduced their direct involvement with the DOD to prioritize commercial markets, enhance profitability, and project a more commercial corporate image. This shift was driven by economic, regulatory, and reputational factors. The simplest reason is that DOD contracts became far more actively managed with draconian limits on profit margin and changes in how costs were associated in accounting systems.

During the Reagan Cold War era, large firms like Lockheed, Boeing, and GE secured lucrative contracts for aircraft, electronics, and weapons systems. However, post-Cold War budget cuts, starting with the 1990s "peace dividend," slashed defense spending. Commercial markets, fueled by globalization and technology booms, offered higher margins and stability.

The complexity of DOD contracts also deterred involvement. The Federal Acquisition Regulation imposed stringent cost-accounting standards, audits, and compliance requirements.

A key strategy for minimizing DOD ties was spinning off or selling defense-focused subsidiaries, allowing companies to isolate low-margin or high-risk military work while preserving a commercial image. In 1993, GM sold its Allison Gas Turbine division to Rolls-Royce, exiting defense to focus on automotive and consumer markets. Similarly, IBM and Ford Motor Company divested their defense-pure business units to Loral Corporation. This move let IBM concentrate on commercial computing, avoiding defense's regulatory scrutiny and cyclical budgets.

GE followed suit, restructuring to reduce defense exposure. In the 1980s, GE's Aerospace unit was a major DOD contractor, producing radar and avionics. By the 1990s, GE sold portions of this unit to Martin Marietta (now part of Lockheed Martin) for $3 billion, redirecting resources to commercial ventures like medical imaging and power systems. This divestiture aligned GE's brand with innovation and consumer technology, distancing it from military associations that could deter foreign institutional investors.

Technology firms, in particular, sought to avoid defense's stigma to attract talent and customers. In the 2000s, Silicon Valley giants like Google and Microsoft faced employee protests over DOD contracts like Project Maven, a drone AI initiative. While these firms maintained some defense work, they downplayed it publicly, emphasizing consumer products like cloud services or search engines.

❖ ❖ ❖

## Set Asides

In the industry, "Set-Aside" categories carve out a slice of the procurement pie for small suppliers. Giving special status to small companies has been officially described as a way to broaden the supply base and help these companies be competitive. The optimistic opinion would be that the supply chain for the US government should be spread geographically and leveling the playing field is a worthy social-economic investment. In January 2025, there were two executive orders addressing a broad governmental change of any approach to the concepts of diversity, equity, and inclusion. The Small Business Administration and General Services Administration are expected to significantly alter the definitions and roles of set-aside programs.

Set-aside categories, as they exist now, target distinct small business types under the FAR. The broadest is the Small Business Set-Aside, for firms with fewer than 500 employees, can cover everything from machine shops to software developers. If a contract is under a certain dollar value and meets some minimum definitions of competitive suppliers, it can be awarded in a much simpler process. There are four other categories used in DOD supplier preferences: Service-Disabled Veteran-Owned Small Businesses (SDVOSBs), Women-Owned Small Businesses (WOSBs), Historically Underutilized Business Zone (HUBZone), and the 8(a) Business Development Program. These special categories attempt to aid supposedly disadvantaged or minority-owned firms in winning government contracts.

This objective is a social-economic effort with supporters claiming it spreads opportunity and engages under-utilized firms. Critics of the set-aside programs argue that government procurement should not be a social investment.

One of the most powerful, and questionable, set-aside categories is the Alaskan Native Corporation (ANC) set-aside, championed by Senator Ted Stevens in the 1980s. Stevens, an Alaskan Republican, saw the 1971 Alaska Native Claims Settlement Act as untapped potential. He crafted an exception in the 1986 Small Business Act, letting ANC firms bypass size caps and compete as "small" despite parent tribal-owned company's size. This aimed to funnel government procurement spend to Alaska, boosting jobs in a remote economy. By 2006, loopholes exploded: ANCs like Chugach Alaska Corporation grew into $1 billion giants, winning sole-source deals like a $2.3 billion Army contract for logistics. Arguably few jobs in Alaska have sprouted from this program, but the program is financially beneficial enough to many programs for it to stay competitive.

The game of certifying small businesses into these categories, and receiving advantages in access to DOD procurement dollars, can be a predatory cottage industry. There are YouTube channels, podcasts, and guidebooks for self-proclaimed experts in government contracting. This is absolutely a situation where individual mileage may vary. There are definitely expert consultants who do help small and non-traditional defense contractors get access to decision makers and engage in this competitive market. Unfortunately, the snake oil sales are a huge part of this niche. For example, as soon as a company registers in the SAM.gov portal there are immediately daily phone calls by vultures promising to help get companies noticed for government contracts. They collect fees from small companies just to do basic processes and administrative tasks.

◈ ◈ ◈

## Silicon Valley

Everyone else in defense contracting is not just small machine shops and specialty manufacturers. The biggest hubs of the digital economy are also invested in the defense economy. Silicon Valley's rise as a launchpad for defense technology research and development traces back to World War II, when a blend of academic talent, military need, and industrial innovation turned a sleepy farming region into a tech powerhouse. Once known for orchards, the area south of San Francisco transformed through Stanford University's leadership, federal funding, and a postwar push for cutting-edge systems. This origin story, rooted in radar, electronics, and Cold War rivalries, set the stage for today's defense startups like Anduril and Palantir, tying the Valley's legacy to the defense industrial enterprise.

The spark ignited in the 1930s at Stanford, where engineering professor Frederick Terman, dubbed the "Father of Silicon Valley," saw potential in radio technology. He mentored students like William Hewlett and David Packard, who founded Hewlett-Packard (HP) in 1939 in a Palo Alto garage. Their early work on audio oscillators caught the Navy's eye, but World War II supercharged the shift. With Japan's attack on Pearl Harbor in 1941, the U.S. raced to counter enemy radar and communications. Terman, tapped by the Army Signal Corps, led Stanford's Radio Research Laboratory, developing countermeasures like jammers. HP and other local firms pivoted to military contracts, crafting electronics for bombers and ships, laying a foundation for defense R&D.

The Space Race and the Cold War fueled growth. The Soviet Union's 1949 atomic bomb test and 1957 Sputnik launch were total catalyst events for US technological prioritization. Stanford Industrial Park, launched in 1951 by Terman, lured companies with cheap land and university ties. By 1956, Lockheed opened its Missile and Space Division in Sunnyvale, drawn by talent and proximity to a naval airbase called Moffett Field. Their Polaris missile program, a $5 billion effort, cemented the Valley's defense role. Federal dollars poured in, with the DOD and NASA funding 60 percent of Stanford's engineering budget by the 1960s.

In 1955, William Shockley, co-inventor of the transistor, founded Shockley Semiconductor in Mountain View. His team, including Robert Noyce and Gordon Moore, split off in 1957 to form the integrated circuit pioneer Fairchild Semiconductor. The Minuteman missile program snapped up Fairchild's chips, with DOD contracts hitting $100 million by 1965. Noyce and Moore split off in 1968 to found Intel. They are both legends in the field and have engineering concepts named after them (Moore's law is famous for predicting the exponential decrease in chip size and increase in power).

The post-9/11 security environment, launch of the Global War on Terror, and the Silicon Valley recovery after the dot-com bust all started to coalesce by 2005. In-Q-Tel, established in 1999 by the CIA, became a catalyst for VC involvement in defense. In-Q-Tel's mission was to identify and invest in cutting-edge technologies that could enhance U.S. intelligence capabilities, bridging the gap between Silicon Valley's innovation ecosystem and the intelligence community. By 2000, In-Q-Tel was actively investing in areas like cybersecurity, data analytics, and geospatial intelligence, with early successes like Keyhole (acquired by Google to become Google Earth).

In-Q-Tel's model was unique: it operated as a non-profit VC firm, providing seed funding and strategic guidance to startups while facilitating connections with government agencies. Its portfolio will have grown to over 300 companies by 2025. Peter Thiel, a co-founder of PayPal and Founders Fund, emerged as a central figure in the defense tech boom. Thiel's libertarian philosophy and belief in technology as a driver of national strength led him to focus on defense-oriented startups. In 2004, he co-founded Palantir Technologies, which became a cornerstone of modern defense tech. Palantir's Gotham platform, designed to integrate and analyze disparate data sources, found immediate applications in counterterrorism and military operations, securing contracts with the Department of Defense and intelligence agencies.

Thiel's influence extended beyond Palantir. Through Founders Fund, he backed other defense unicorns, including Anduril. Thiel's vision emphasized the need for the U.S. to counter China's technological rise, advocating for closer collaboration between Silicon Valley and the Pentagon.

In 2015, the DOD launched the Defense Innovation Unit Experimental (DIUx) to accelerate the adoption of commercial technologies. Headquartered in Silicon Valley, DIUx aimed to overcome the cultural and bureaucratic barriers that hindered collaboration between the Pentagon and tech startups. Under Secretary of Defense Ash Carter, DIUx focused on rapid prototyping and contracting, attempting to offer a streamlined pathway for VC-backed companies to secure DOD contracts.

DIUx's early years were rocky, with criticism over unclear objectives and limited funding. By 2018, it had evolved into the Defense Innovation Unit (DIU), with expanded authority and a broader mandate. The DIU pioneered initiatives like the Commercial Solutions Opening, enabling startups to pitch solutions directly to the DOD. By 2025, DIU had facilitated contracts for technologies ranging from AI-driven logistics to autonomous drones, with companies like Anduril and Shield AI benefiting from its streamlined processes. The founding director of DIU, Raj M. Shah, co-authored his own story of the DIU origin, challenges, and then success in his 2024 book *Unit X: How the Pentagon and Silicon Valley Are Transforming the Future of War*. Mr. Shah is a managing partner at Shield Capital which invests in the dual-use defense tech startup world.

The rise of VC in defense has produced several unicorns that are trying to challenge the "legacy" defense industry. Palantir, valued at over $60 billion by 2025, transformed how the DOD and intelligence agencies process data. Its software has enabled real-time insights for operations like tracking insurgents in Afghanistan, earning it a reputation as a game-changer. The CEO of Palantir, Dr. Alexander Karp, released his full-throated call to action for Silicon Valley to get off the sidelines and engaged in the battle for Western influence in the world. *The Technologic Republic* was an immediate best seller at the start of 2025 and describes the Palantir approach to national security through technology as a "first principles" imperative. The Palantir CTO, Shyam Sankar, is an active voice for the role of technologists in a "First Breakfast" to recover from the "Last Supper".

Anduril Industries, founded in 2017 by Palmer Luckey, is a Silicon Valley-based defense technology company disrupting traditional defense contracting with innovative, AI-driven solutions. Valued at $14 billion by 2025, Anduril has emerged as a unicorn backed by venture capital firms like Peter Thiel's Founders Fund and Andreessen Horowitz. Anduril's flagship product, Lattice, is an AI-powered operating system that integrates sensors, drones, and cameras to provide real-time battlefield awareness. Lattice enables autonomous coordination of assets, such as Anduril's Ghost drones, which are used for surveillance and reconnaissance. The company also develops advanced hardware like the Roadrunner, a modular, AI-guided missile for counter-drone defense, and the Altius-600, a loitering munition deployed in Ukraine. These systems prioritize rapid deployment and adaptability, challenging the slow, expensive models of traditional defense primes.

The major problem is that the Department of Defense is not good at buying software. The "Valley of Death" in defense technology is the critical gap between the development of a promising prototype and its successful transition to full-scale production and deployment within the U.S. military. All of the steps and processes of the Defense Acquisition

Lifecycle were established long before SaaS, Agile, or GitHub. VC-backed firms are experimenting real time in merging the legacy defense industry game of former military officers, K Street lobbyists, and finding ways to actually sell their technology to the DOD.

The primary obstacle in the Valley of Death is the DOD's lengthy and risk-averse procurement system. While startups excel at rapid prototyping, often delivering AI-driven drones or cybersecurity solutions in months, scaling these technologies requires navigating multi-year contracts, rigorous certifications, and competition with established primes. Funding dries up as venture capital, designed for quick returns, struggles to sustain companies through extended DOD timelines. There are countless attempts at pitch days, innovation hubs, and other micro-paths that try to go from a SBIR contract to the promised land of a program of record. But even Palantir had to sue the Army to stop them from developing a legacy government software and hardware tool that was just trying to copy what Palantir had already launched. Bureaucratic resistance, or "bureaucratic antibodies," as described in *Unit X*, further complicates the transition. Congressional oversight, shifting priorities, and entrenched interests favor legacy systems over unproven innovations.

The TL;DR for defense in Silicon Valley is that the DOD and some significant founder personalities have deep history in bringing technology to the defense of the Western world. There is currently significant funding and extremely intelligent people trying to get further involved and change the technology landscape for defense. The defense startup scene has a significant mountain to climb considering they do not have sturdy defense voices representing them in the

Congress and Senate. Fighting for a share of the defense industrial base is not based only on a simple "best technology wins." The decisions cross many stakeholders and every defense dollar still has to be granted by Congress.

❖ ❖ ❖

## Private Equity and Capital

Startups and venture capital are at one end of the flow of capital into defense. The biggest shaper of the players in defense though come from private equity. PE has become a significant force in the U.S. defense industrial base, stepping into a landscape once ruled by giant prime contractors like Lockheed Martin and Boeing. These firms, wielding billions in capital, acquire, reshape, and sell defense-related companies at a scale unmatched in most other sectors. Unlike venture capital's focus on nascent startups or primes' long-term production cycles, PE targets established small to mid-sized firms, aiming for quick value creation through operational tweaks, strategic mergers, or market repositioning. With the DOD's $842 billion FY 2025 budget, PE's role has grown, managing roughly $5.6 billion in defense investments in 2023 alone.

Private equity in the defense business is still a hard sector that attracts bold investors. One of the pre-eminent players has been the Carlyle Group, a private equity giant that's been a heavyweight in the U.S. defense industry for decades. Founded in 1987 in Washington, D.C., Carlyle carved out a unique identity by leveraging political connections and strategic investments, particularly in defense contracting. Dan

Briody's 2003 book *The Iron Triangle: Inside the Secret World of the Carlyle Group,* reads like a thriller. The book tells a dramatic story full of high-stakes deals, influential figures, and whispers of cronyism.

Carlyle's rise wasn't accidental. Unlike traditional buyout firms in New York or Chicago, Carlyle set up shop in D.C. As one investor noted, "foreigners have to come to Washington" (Briody, 2003). This proximity to power gave Carlyle unparalleled access to policymakers and defense insiders. By the early 2000s, it managed over $14 billion, with a portfolio spanning defense, aerospace, and telecom. Its defense investments included companies like United Defense, later sold to BAE, a project Carlyle fought to save despite Pentagon skepticism. This wasn't just business; it was a masterclass in influence, with former Secretary of Defense Frank Carlucci steering the ship. Carlyle's roster read like a who's who of global elites: George H.W. Bush, James Baker III, and even the Bin Laden family had stakes until public scrutiny forced a split post-9/11.

Briody's *The Iron Triangle* pulls no punches. The book argues that Carlyle epitomizes the "military-industrial complex" President Eisenhower warned about in 1961. He details how Carlyle's executives, many fresh from high-level government posts, used their connections to secure contracts and sway policy. Telling a sensationalist story can be entertaining, but Carlyle and other private equity players do provide a consistent influence that keeps the sector strong and churning for better alignment.

Carlyle's story isn't just about one firm; it's a window into the defense industry's evolution. By the 2000s, traditional contractors like Lockheed Martin faced competition from agile, VC-backed startups, and then at a rising tide of smaller firms using the set-aside advantages. Private equity (and hedge fund involvement in defense) shows that strength lies in leveraging relationships, a reminder that defense isn't just about tech but about who you know. This dynamic makes the Valley of Death, that perilous gap between prototype and deployment, even trickier for newcomers.

Critics argue that a profit motive and the revolving door between government and industry, which Carlyle mastered, are unethical. But the defense industry is no different from other large, regulated sectors, there is a small cohort of people knowledgeable about the field and the game is only played between the companies and the lone government customer. Still, Carlyle's scale, with billions in defense assets, amplifies the stakes. By 2025, private equity and venture capital will be more intertwined. The defense tech segment reflects frequent mergers and acquisitions and the acqui-hire model where a larger firm buys a company for a few key hires. Because defense isn't just about tanks or drones; it's about who controls the purse strings and the policies.

Arlington Capital Partners, also D.C.-based and founded in 1999, is heavily focused in defense investments. ACP has over $9.1 billion in assets under management in 2025 and was recently successful in selling its rollup star BlueHalo to AeroVironment. It builds "platform" companies by merging smaller players into cohesive units, often selling to primes or strategics. In 2024, Arlington formed Keel, blending Pegasus Steel, Merrill Technologies Group, and Metal Trades into a $1 billion-plus fabricator for Navy submarines. Its 2016 creation of Polaris Alpha, merging three portfolio firms, grew into a cyber and intelligence provider, sold to Parsons in 2018 for $489 million. Arlington's 2007 Novetta build, merging White Oak and FGM is a regularly cited deal that proves the ACP model. With $3.8 billion Fund VI closed in 2024, Arlington keeps betting on defense growth, leveraging hands-on management and industry know-how.

Both firms enhance the DIB differently. Carlyle's scale and global reach, with 2,300 staff across 29 offices, let it tackle big bets like Booz Allen, driving broad capability jumps. Arlington's leaner, sector-specific focus crafts tailored solutions, like Keel's submarine structures. They boost small suppliers, vital to the 1 million-plus employees making up the defense industrial base workforce, by injecting capital lower in the value stream.

PE's role sparks debate. Proponents say it modernizes a creaky DIB. Critics, including GAO, flag risks: debt-heavy buyouts, like TransDigm's $20 billion post-PE load threaten solvency. Also the short hold periods (Carlyle's five-year ARINC flip) may skimp on R&D for quick exits. Consolidation can shrink competition, echoing Orbital ATK's 2018 sale to Northrop Grumman which has effectively ended competition in defense heavy weapons and ammunition.

❖ ❖ ❖

## Think Tanks

Think tanks play a pivotal role in the defense industrial base serving as intellectual engines that shape policy, strategy, and innovation. Unlike Venture or Private Equity, they don't show up with funding, they join the conversation through academic papers and plenary sessions that get quoted and spread from the halls of Congress to the industry trade shows. These organizations, blending analysis with advocacy or research, influence everything from acquisition priorities to technological development, acting as bridges between government, industry, and academia. Policy-focused groups like the Center for Strategic and International Studies (CSIS) and the American Enterprise Institute (AEI) steer national security debates from the "outside", while government-funded research entities like MITRE and RAND also heavily influence the DC acquisition bubble and talking heads. In July 2025, Def. Secretary Hegseth issued a bizarre ban on DOD staff participating in think tank events.

CSIS, founded in 1962 at Georgetown University and now independent, ranks among the top defense think tanks, known for its centrist, pragmatic take on security. With a $50 million annual budget in 2023, partly from DOD grants and industry donations, CSIS churns out reports like its 2022 study on hypersonic weapons, urging faster acquisition to counter China. Its Defense-Industrial Initiatives Group tracks DIB health, flagging supply chain risks during Ukraine aid surges. CSIS hosts heavyweights like ex-Defense Secretary Mark Esper, amplifying its voice in Congress and the Pentagon. Its 2023 "Rebuilding the Arsenal" series pushed for industrial base resilience, influencing FY 2025 POMs with calls for $20 billion in manufacturing upgrades.

AEI, launched in 1938, brings a conservative lens to the DIB, rooted in free-market principles. With a $40 million budget in 2023, funded by corporate donors, AEI critiques government overreach and champions private sector solutions. Its 2021 report on defense spending, penned by scholar Mackenzie Eaglen, argued for a $900 billion budget to outpace China, which arguably provided talking points to defense hawks during the NDAA process. AEI's focus on great power competition, like its 2023 AI warfare papers, nudges DOD toward tech investments, indirectly boosting primes and startups. Less technical than CSIS, AEI excels at ideological framing, hosting figures like Newt Gingrich to rally political will, though its partisan bent can limit bipartisan reach.

MITRE was established in 1958 as an FFRDC spin-out from MIT. With $2.5 billion in 2023 federal funding, MITRE provides systems engineering to research efforts and actual program staff. MITRE's 9,000 staff, heavy on PhDs, bridge acquisition phases, advising PMs during different phases of the lifecycle. Its neutrality, mandated by FFRDC rules, keeps it free of industry profit motives, although The Professional Services Council and contractors know FFRDCs are unfair competiton.

RAND Corporation, the first FFRDC, born in 1947 from an Army Air Forces deal with Douglas Aircraft, blends research with strategy. Its $400 million 2023 budget, split between DOD, Air Force, and other sponsors, funds 1,800 researchers tackling long-term questions. RAND's 1960s systems analysis shaped Cold War deterrence, while its 2022 wargames about the Taiwan conflict informed force posture shifts. It advises on Acquisition Lifecycle gaps, like drone scalability for Replicator, and runs Project Air Force, a $70 million think tank within a think tank. RAND's independence lets it critique DOD flops, like the Littoral Combat Ship, though its academic pace can lag urgent needs.

Policy think tanks like CSIS and AEI influence the DIB indirectly, setting the strategic table. CSIS's data-driven push for industrial capacity sways service program priorities, while AEI's hawkish stance fuels budget discussions in a GOP-controlled House. They engage Congress via testimony, like CSIS's 2024 NDAA briefings, and host industry forums, aligning DIB priorities with threats. Research FFRDCs like MITRE and RAND dig into execution, supporting PMs with tech specs during Engineering and Manufacturing Development or modeling sustainment costs in Operations and Support.

The Center for a New American Security (CNAS), founded in 2007 by Michèle Flournoy and Kurt Campbell is a credible upstart in the think tank field. They inform U.S. national security policy with a mission to deliver "strong, pragmatic, and principled" solutions. Emerging amid the Iraq and Afghanistan wars, CNAS blends bipartisan research, policy engagement, and leadership development to address global challenges like China's rise, cyber threats, and great power competition. Its stated purpose is to craft innovative, fact-based strategies that inform policymakers, nurture future leaders, and engage the public, ensuring America's security in a turbulent world. CNAS's influence through reports like *Rising to the China Challenge* (2020), has driven U.S.-Taiwan policy. They tend to be considered a credible voice from the left, which is generally not accepted as a strong stakeholder in the current defense realm.

These organizations provide valuable analysis and steer public conversation. While most of the public is not actually paying attention to such wonky policy details, the defense industry is a constantly moving, full-contact sport. If every stakeholder in the process is completely happy, then that means that hard decisions have not been made, and the technology and capacity are most likely not providing a significant advantage for the current and future battlefields.

❖ ❖ ❖

## Foreign Companies

While the US defense industry is definitely a home field game, international companies like Germany's Rheinmetall, are competing in the market in a major way. The current administration is shocking the status quo with reactive tariff policy and a very different approach to foreign trade policy. Foreign companies have found paths into the US market through acquisitions (and even old-fashioned decisions like the US buying the best technology in the world) even when it isn't homegrown. International companies like Germany's Rheinmetall, Italy's Leonardo DRS, Britain's BAE Systems, South Korea's Hanwha, Israel's Elbit Systems and Rafael have become major players. These firms aren't just dipping their toes in the Pentagon's waters, they're reshaping how America arms itself, bringing global expertise to a market once dominated by domestic giants. In previous administrations, co-development and unity of NATO materiel was a variable in foreign relations, a lever to be pulled to align allies.

The U.S. defense industry has always been a fortress of innovation, but it's not an island. As threats evolve, like China's hypersonic missiles or Russia's cyber brigades, the Pentagon's hunger for cutting-edge tech has had to evolve to a flattening Earth in the distribution of technical capability. International companies often team up with U.S. firms to navigate the Pentagon's intricate procurement rules.

Rheinmetall, a German powerhouse known for tanks and artillery, has a deep legacy in the crown jewel M1 Abrams program. The M256 main gun on the Abrams was designed by Rheinmetall and then licensed by the US for decades. It started as a NATO aligned effort, but protectionists like New York Representative Samuel Stratton and Senator Chuck Schumer have used amendments and budget control to make sure Rheinmetall has not been allowed to supply tank barrels for the gun system they designed. They've made a major push into the US with a contract for the U.S. Army's XM30 Mechanized Infantry Combat Vehicle program. Their Lynx platform, a next-gen infantry fighting vehicle, is a contender to replace the aging Bradley. They are the biggest provider of artillery ammunition to Ukraine, and they are arguably the strongest prime in position to rearm Europe as the EU must react to shifts in American foreign policy. Lockheed Martin is working with them in Europe for F-35 fuselages as an attempt to keep that platform sold internationally.

Leonardo DRS, the U.S. arm of Italy's Leonardo S.p.A., is another heavy hitter. Based in Virginia, they're a go-to for electronics and sensors, winning contracts for the Army's Short-Range Air Defense System on Stryker vehicles. They've also delivered Trophy active protection systems for

M1 Abrams tanks alongside Israel's Rafael. Leonardo's stock surged 13% in early 2025, reflecting investor confidence in their U.S. defense foothold.

BAE Systems, Britain's defense juggernaut, is practically a household name in U.S. military circles. They're behind the Amphibious Combat Vehicle for the Marines (built with Italy's Iveco) and they bought their way into the Bradley when they acquired United Defense. BAE's Hägglunds CV90, a versatile infantry fighting vehicle, has racked up orders across Europe, and its U.S. division pitches similar tech for Army programs. Their stock climbed 16% in March 2025, buoyed by global defense spending spikes. BAE's edge lies in blending British engineering with American manufacturing, ensuring jobs and tech stay stateside, a political necessity.

In York, Pennsylvania, the BAE Systems plant is more than a factory, it's the lifeblood of a community. They build the Bradley Fighting Vehicle and Armored Multi-Purpose Vehicle, supporting hundreds of jobs. Rep. Jack Murtha, a former Marine from Pennsylvania's 12th District, made it his mission to keep this plant humming. As chairman of the House Defense Appropriations Subcommittee, Murtha used his clout to ensure funding lines for BAE programs were safe and growing. His mastery of earmarks ensured York's economic survival, with The Hill noting he nabbed $150 million in "pork" for his district in 2009. For Murtha, a Vietnam vet, this wasn't just politics; it was personal, a way to protect a region hit hard by the decline of coal and steel.

Murtha's methods were old-school and not without critics. He was accused of strong-arming colleagues and was one of the rare defense hawks in the Democrat Caucus. His ties to BAE ran deep, and rivals cried foul, pointing to Murtha's links with the PMA Group, a lobbying firm that scored $28.1 million in earmarks from him in 2008. The FBI even raided PMA, though Murtha dodged charges. Critics saw this as "pay-to-play," but in York, workers saw a hero who kept the plant off the 2005 Base Realignment and Closure chopping block.

By Murtha's death in 2010, his legacy in York was rock-solid. The BAE plant, still modernizing Bradleys and building Amphibious Combat Vehicles, thrives because of his tenacity. His annual "Showcase for Commerce" in Johnstown, dubbed "MurthaFest," was a love letter to the region's defense industry, with BAE front and center.

South Korea's Hanwha Defense is the new kid on the block, but they're making waves. Known for howitzers and armored vehicles, Hanwha's eyeing U.S. programs like self-propelled artillery to replace the M109 Paladin. Their stock soared 16.7% in early 2025, driven by plans to double howitzer production. Rumors swirl of a partnership with Textron to crack the U.S. market, a smart move given the Army's modernization push. They are growing as an industrial conglomerate with their purchase of the Philly Shipyard, aligning with the calls for more shipbuilding.

This global collaboration isn't without critics. Some worry about dependency on foreign tech, especially from countries like South Korea or Israel, where geopolitical risks loom. Think tanks like CSIS flag the need for a robust U.S. industrial base to avoid over-reliance on an international supply chain. The reality is pragmatic, global threats and localized specialization move faster than domestic R&D can fund. Partnerships with firms like Leonardo or Hanwha let the U.S. leapfrog development cycles, getting tech to the battlefield quicker.

By 2025, these companies are riding a wave of rising defense budgets, with Europe's military spending jump fueling stock gains for BAE, Rheinmetall, and Leonardo. Leonardo DRS, BAE, Hanwha, Elbit, and a growing list of foreign companies with special security agreements approved by the Defense Security Cooperation Agency are competing for a growing slice of the market. Their contributions range from supplying critical components to co-developing major systems, navigating strict regulations like ITAR and Buy American laws. The US Government has had a conflicted policy with international suppliers for decades.

The AUKUS pact, signed in 2021 by the U.S., UK, and Australia, opens new doors. Designed to counter China, it splits into Pillar I (nuclear subs for Australia) and Pillar II (tech sharing in AI, cyber, and hypersonics). Rheinmetall, though not a core member, taps Pillar II via NATO ties and UK collaborations, like its 2024 Boxer vehicle deal with BAE Systems for British forces, eyed for Australian export. Its Virginia-based AUKUS office pitches hypersonic munitions, leveraging German precision to join a $10 billion R&D pool. AUKUS's open architecture lets Rheinmetall integrate with U.S. primes like Raytheon, boosting interoperability.

# CHAPTER 6

# Beans, Bullets, and IBM

## Products, Services, and Contracts

*"The cost of one modern heavy bomber is this: a modern brick school in more than 30 cities. It consists of two electric power plants, each serving a town of approximately 60,000 population. It is two fine, fully equipped hospitals. It is some 50 miles of concrete highway."*
—President Dwight Eisenhower

THE U.S. DEPARTMENT OF DEFENSE operates on a simple yet profound principle encapsulated in the phrase "Beans and Bullets," a nod to the essentials that keep its warfighters fed, armed, and ready. Ever wonder how the U.S. military keeps its troops fed, armed, and ready while maintaining ancient warhorses like the B-52 Stratofortress? The Department of Defense's procurement of "beans and bullets" is a logistical apparatus rivaling Amazon. It is a sprawling operation that keeps soldiers sustained and platforms flying. B-52s, first flown in 1952, still roar through the skies in 2025. It's a story of bureaucracy, ingenuity, and an industry that's as much about spare parts as it is about cutting-edge tech. A famous quote from General Omar Bradley is "amateurs talk strategy, professionals talk logistics."

The phrase "beans and bullets" captures the gritty essentials of military logistics. Beans mean sustenance, MREs (Meals, Ready-to-Eat), or bulk rations for mess halls. Bullets cover ammunition, from 5.56mm rounds for M4 rifles to 155mm artillery shells. The Defense Logistics Agency (DLA) has one of the most ambitious charters in the DOD and across the defense industry. They manage a $40 billion supply chain that delivers everything from freeze-dried beef to Hellfire missiles. In 2024, the DLA awarded contracts worth $3.2 billion to companies like Sysco for food and Winchester for ammo. These deals aren't flashy, seldom making headlines outside of their little niche market, ensuring troops in places like Djibouti or Guam never run dry. The process leans on competitive bidding, but there is no straight-line path to revenue for vendors. Unfortunately, the cottage industry of "consultants" trying to sell online courses about how small businesses can make millions plagues this end of the supply chain.

After consumables, there are the major weapons platforms, like the B-52. This eight-engine behemoth, built by Boeing, is a flying testament to durability. Designed to drop nukes during the Cold War, it's been retooled for precision strikes, carrying Joint Direct Attack Munitions (JDAMs) over Afghanistan or Iraq. The Air Force plans to keep B-52s airborne until 2050, nearly a century after their debut. That is either exceptional ingenuity and value for the taxpayer, or potentially an embarrassment that our acquisition system seems to have frozen during the Cold War (sorry, couldn't pass up the pun). The B-52 is an even more amazing story considering it was allegedly designed over one weekend in a hotel in Ohio. But how does an airframe survive and thrive for 100 years? Through constant upgrades and a supply chain that's part scavenger hunt, part high-tech hustle. The B-52's TF33 engines, for instance, haven't been made since the 1980s, so the Air Force relies on cannibalization, stripping parts from retired airframes, and reverse-engineering components. In 2023, Boeing inked a $2.6 billion contract to modernize B-52s with new Rolls-Royce F130 engines, radar, and avionics.

The supply chain for these platforms is a beast of its own. Companies like Collins Aerospace and Honeywell supply avionics and sensors, while smaller firms like Moog produce niche parts like hydraulic actuators. The DOD's Defense Production Act investments ($500 million in 2024) attempt to bolster domestic manufacturing for critical components. The supply chain from the main street machine shops up through the primes is a shrinking pool. A single broken widget, like a B-52 landing gear strut, can ground a plane if the original manufacturer went out of business years ago. The DLA's "diminishing manufacturing sources" program tackles this, contracting firms to recreate obsolete parts using 3D printing or modern alloys. It's like keeping a classic car running with custom-machined bolts.

This ties into the broader defense industry we've been exploring. Think of the B-52's longevity as a cousin to Watervliet Arsenal's cannon-making, both preserved by congressional champions like Rep. Jack Murtha or the Stratton Amendment. The supply chain's complexity mirrors the "iron triangle" dynamics we saw with Carlyle Group, where industry, government, and politics intertwine.

Foreign players, like BAE Systems supplying M2 Bradley parts or Rheinmetall's 155mm shells, add another layer. But global reliance has risks, Ukraine's ammo shortages in 2024 showed how fast stockpiles dwindle when supply chains stretch thin. Then, add a complete reimagining of the global trade concept in America, and this mission of keeping hardware ready to fight becomes extremely difficult.

Maintaining these platforms isn't cheap. The B-52's annual sustainment costs run about $1.2 billion, a fraction of the $15 billion for a single new B-21 Raider. That's the trade-off: refurbishing old platforms saves cash but leans on a creaky supply chain. The DOD's answer is public-private partnerships, like Boeing's Oklahoma City depot, where Air Force mechanics and contractor techs overhaul B-52s side by side. These setups cut downtime, but they're not foolproof. A 2022 audit found 30% of B-52 spare parts were back ordered, forcing workarounds like borrowing from allied nations' stocks.

Let's not forget the human side. The folks at places like Tinker Air Force Base, where B-52s get their facelifts, are unsung heroes. They're machinists, logisticians, and engineers keeping planes older than their grandparents flying. Their work echoes the grit of York's BAE workers, who Murtha fought for, or the Watervliet crew forging barrels. It's a reminder that defense isn't just about shiny new drones from Silicon Valley startups, sometimes, it's about coaxing life from Cold War relics.

The DOD's procurement of beans and bullets, paired with the supply chain for long-lived platforms, is a masterclass in juggling the mundane and the monumental. MREs and ammo keep troops going; spare parts keep B-52s soaring. But it's a tightrope walk. Global disruptions, like semiconductor shortages or geopolitical flare-ups, can choke the flow. The Pentagon's push for "resilient logistics" aims to shore up weak spots. Think tanks like CSIS warn that China's supply chain dominance in rare earths for electronics poses an existential threat.

The U.S. is tackling China's chokehold on rare earth minerals. On April 24, 2025, President Trump signed an Executive Order, "Unleashing America's Offshore Critical Minerals and Resources," to kickstart deep-sea mining for nickel, cobalt, and rare earths. These materials are critical for chips, missiles, and batteries. With China dominating 90% of rare earth magnets, it's a gutsy move to cut foreign reliance. The EO tells the Commerce Secretary to speed up permits under the 1980 Deep Seabed Hard Mineral Resources Act, clearing the way for firms to mine seabed nodules. NOAA is tasked with mapping ocean floors, and a 60-day deadline aims to streamline licensing. The Pentagon and Energy Department will tap the National Defense Stockpile, linking to DOD supply chains.

So, what's the big picture? The DOD's ability to procure beans and bullets and sustain platforms like the B-52 is a testament to its logistical muscle, but it's not invincible. It's a patchwork of contracts, partnerships, and sheer stubbornness, knitting together giants like Boeing with small-town shops. The B-52's century-long run isn't just about engineering, it's about a supply chain that refuses to quit, even when the odds say otherwise. In a world of flashier tech, there's something reassuring about that old bomber still taking flight, fueled by the same dogged system that keeps soldiers fed and armed.

❖ ❖ ❖

## A Lifetime or Two of Hardware

Ever wonder why the U.S. military still flies bombers older than most of its pilots or why new weapon systems seem to balloon in cost and complexity? J. Ronald Fox, a Harvard Business School professor and defense acquisition expert, tackled these questions in his books *Arming America* (1974) and *The Defense Management Challenge* (1988). Together, they paint a vivid picture of how the defense industry molds major weapon system platforms and why the system keeps old platforms chugging along instead of replacing them. It's a system of industry clout, bureaucratic inertia, and incentives that sometimes prioritize profit over progress, and it's as relevant in 2025 as it was decades ago.

The defense industry, as Fox describes in *Arming America*, isn't just a supplier; it's a sculptor, chiseling platforms to fit a tangled web of military needs, political pressures, and corporate ambitions. Arguably, the DOD embracing the concept of "Modular Open System Architectures" is a way to make pieces and components more replaceable for the purpose of keeping vehicles and ships in operation longer. Critics have argued for decades that the acquisition system incentivizes prime contractors to "Gold Plate" their systems, promising performance thresholds way above the base requirements. But if only a few platforms a decade are fielded, then there are very few at-bats for primes to capture market share and revenue. In *The Defense Management Challenge*, he notes that firms often lowball initial bids to win contracts, knowing they can recoup losses through upgrades or spares. This shapes platforms into complex beasts, like the Navy's Arleigh Burke destroyers, packed with Aegis systems that demand constant tweaking. It's like building a Swiss Army knife where every blade costs a fortune and needs its own maintenance crew.

But here's where it gets messy: the industry's influence doesn't stop at design. Fox argues that contractors have a vested interest in making platforms indispensable. In *Arming America* he details how firms lobby for "sole-source" contracts, ensuring they're the only ones who can maintain or upgrade systems like the F-16. This locks the Pentagon into long-term relationships. The defense industry is forced to engage with Congress since they are stakeholders in the checkbook. But then critics accuse them of being too cozy and making decisions based on revenue stability and not battlefield capability. Platforms are often built with jobs in mind, not just combat. Fox points out that a single platform, like the C-130 Hercules, can support thousands of jobs across states, making cancellation political poison. It's a machine that runs on economic fuel as much as a strategic need.

The system incentivizes stretching platform lifespans over building anew. Upgrading a B-52 with new engines or avionics costs a fraction of the development costs of an entirely new airframe, and the industrial base provides hundreds of millions of screws, bolts, and wires. Fox notes that the Pentagon's budget structure rewards "incremental improvements" over risky new designs, since Congress balks at blank checks for unproven tech. The F-15, for instance, has morphed into the EX variant, keeping assembly lines alive in St. Louis. It's like patching an old jacket instead of buying a new one, it's cheaper, safer, and keeps the tailor employed.

This ties into the supply chain dynamics of "beans and bullets." The Department of Defense's nightmare scenario is the creeping death of diminishing manufacturing sources. A 2022 GAO report slams the DOD's DMSMS program for tracking these risks but fumbling solutions, leaving platforms vulnerable. With shrinking suppliers in the defense industrial base, the DOD faces threats from foreign reliance and a shrinking pool of domestic manufacturers, down from 51 primes in the 1990s to under 10 today.

The DOD's 2024 National Defense Industrial Strategy (NDIS) is a nod in the right direction, pushing for tougher supply chains and faster buys to keep legacy systems like the Bradley operational. Commissions like the Manufacturing Capability Expansion and Investment Prioritization (MCEIP) office are throwing cash at the problem, but $17.5 million for microelectronics and $20 million for manganese are less than most car dealerships carry in inventory on the lot.

Despite the growing supply chain catastrophe, keeping old platforms alive isn't always a bad deal. The B-52's adaptability proves Fox's point that versatile designs can outlast their makers' wildest dreams. New threats, like China's hypersonics, demand fresh platforms, but the system's inertia, dependent on industry lobbying and Pentagon process enforcement, slows the pivot. Fox's books, though dated, nail this tension: the defense industry builds platforms to last, but the system's rigged to keep them going longer than makes sense.

◆ ◆ ◆

## Services

The Abrams Doctrine, named after General Creighton Abrams, fundamentally changed U.S. military logistics in the 1970s after Vietnam by shifting support services from Active Duty forces to a combination of Reserve, National Guard, and civilian contractors. As Army Chief of Staff from 1972 to 1974, Abrams sought to avoid another unpopular draft by tying combat power to broader societal buy-in. He deliberately moved critical support functions like transportation, maintenance, and supply out of the active ranks, ensuring wars required mobilizing reserves (or outsourcing to civilians), thus forcing public and Congressional consent. This doctrine, formalized after his 1974 death, laid groundwork for the Global War on Terror (GWOT) concept of sustainment in combat. "Services" contracting, especially via the Logistics Civil Augmentation Program (LOGCAP), ballooned to meet sprawling demands.

Abrams' vision stemmed from Vietnam's fallout, where a 2.5 million-strong Active Duty force stretched thin, relying on draftees. By 1973, with the all-volunteer force, he cut active strength to 785,000, offloading 40 percent of logistics functions to 870,000 Reservists and National Guard. Units like the 82nd Airborne lost organic truck battalions, banking on Reserve activation (and eventually private firms) for beans-and-bullets sustainment. The 1991 Gulf War tested this, with 140,000 reservists and Kellogg Brown & Root (KBR) hauling fuel under a $200 million LOGCAP I deal, proving the model. By GWOT's 2001 kickoff, Active Duty stood at 1.4 million, but Abrams' legacy meant contractors filled gaps, a shift Congress backed with hundreds of millions in "Overseas Contingency Operations" funding.

LOGCAP's explosion during GWOT turned this doctrine into a contracting juggernaut. Launched in 1985 to outsource logistics, LOGCAP facilities, maintenance, and supply absorbed larger and larger sections of the budget. By 2008, 160,000 contractors dwarfed 140,000 troops in Iraq, building bases, feeding soldiers, and trucking supplies across war zones. Costs soared, with $5 billion yearly peaks, driven by urgent needs the lean Active Duty force couldn't meet. Estimates of this contingency funding across GWOT are $600 billion on the low side and over $1 trillion on the high side. There are differing opinions on which costs should be included in the estimate. LOGCAP V, active in 2025 with $10 billion tasked to Fluor and others, sustains 50,000 troops globally, from Qatar to Poland. Congress fueled this, appropriating $15 billion in 2003 supplemental funds alone, sidestepping PPBE delays.

Beyond LOGCAP, GWOT birthed a "Services" contracting boom across all types. Systems Engineering and Technical Assistance (SETA) contracts surged, staffing Pentagon offices with expertise Active Duty shed. Firms like Booz Allen Hamilton, under a $2 billion 2023 deal, embed 10,000 analysts in acquisition. SETA's $20 billion annual slice by 2025 reflects Abrams' outsourcing ripple, with Congress pushing $5 billion in 2023 NDAA tweaks for tech acceleration. These contracts, often Cost-Reimbursement for flexibility, fill gaps in uniformed engineers, though critics note overreliance risks institutional knowledge loss.

Consulting services ballooned too, advising strategy and efficiency. Deloitte's $300 million 2023 audit contract, a Time-and-Materials deal, tracks DOD's $3.5 trillion assets. Although the DOD is infamous for "never passing an audit," but realistically a system of this magnitude could never be fully itemized. McKinsey's $1 billion in 2023 work revamps logistics, echoing Abrams' shift, with as many as 5,000 consultants. Congress drives this, mandating $500 million in 2022 for management reviews, despite DOD grumbling over cost, ensuring civilian firms steer policy where Active Duty once did.

Leidos' $7 billion 2023 DES contract runs DOD networks, a Fixed-Price deal for 50,000 users. Palantir's $1.5 billion 2023 AI analytics targets enemies, a leap from Vietnam's paper maps. A question that is being raised by the new administration is if $50 billion in IT contracting spend and contractors like CACI with 10,000 staff, are really a Fifth Estate extension of the US Government.

❖ ❖ ❖

## The Suits Meet the Uniforms

The story of consulting in the DOD and the defense industry kicked off in the mid-20th century, a time when brainy number-crunchers began reshaping how America fights. Robert Sobel's *The Whiz Kids: The Founding Fathers of American Business* describe the starting era as Secretary of Defense McNamara brought the latest in management theory from Ford to the Pentagon. Walt Bogdanich and Michael Forsythe's *When McKinsey Comes to Town* provides a critical expose of ambition, influence, and a system hooked on outside expertise. From legacy players like IBM to the "Big Three": McKinsey, Bain, and BCG. Consulting is where the suits meet the uniforms.

The seeds of DOD consulting sprouted after World War II, when the military needed sharper ways to manage its ballooning budgets and complex systems. Sobel's *The Whiz Kids* chronicles the rise of Robert McNamara and his Ford Motor Company crew, who brought statistical wizardry to the Pentagon in the 1960s. These "Whiz Kids," many ex-Army Air Forces analysts, pushed systems analysis, crunching data to optimize everything from B-52 deployments to troop logistics. Their approach caught the eye of firms like McKinsey, which was already an Ivy League to corporate boardroom funnel. McKinsey was dipping its toes in DOD waters, helping streamline acquisitions for platforms like the F-4 Phantom. IBM, meanwhile, was less about strategy and more about tech muscle, installing early mainframes in the Pentagon. It was like giving the military a shiny new calculator to track its "beans and bullets."

As an aside, the first "computers" installed at the Pentagon was the UNIVAC I, built by Remington Rand, arriving in 1952. It was followed by IBM's 701 "Defense Calculator" in 1953. Priced at $1 million, the UNIVAC I was a beast, installed in the Pentagon's basement, A Ring, Level B, in climate-controlled rooms to protect its vacuum tubes. The IBM 701 joined it there, tucked away for security and space.

The UNIVAC I, with 12 KB of mercury delay-line memory, crunched Air Force logistics estimates at 1,900 operations per second. The IBM 701, with 72 KB, tackled ballistics and nuclear simulations, churning 16,000 additions per second. Both powered Korean War logistics and Cold War planning. These basement beasts were guarded tightly and laid the DOD's digital foundation, echoing the "Whiz Kids" data-driven shift. They also hinted at supply chain woes, IBM's proprietary parts foreshadowed diminishing manufacturing issues.

As the Cold War heated up, the Big Three (McKinsey, BCG, and Bain) saw dollar signs in defense. McKinsey, with its buttoned-up prestige, became a DOD darling. Advising on everything from missile programs to base closures. Bogdanich and Forsythe note McKinsey's knack for embedding itself in government. BCG, founded in 1963, brought its growth-share matrix to defense, helping prioritize investments in systems like the M1 Abrams. Bain focused on cost-cutting in Navy shipbuilding in the 1980s. Their influence grew as defense budgets swelled, with contractors like Boeing leaning on consultants to navigate Pentagon red tape.

The defense industry's reliance on consultants mirrored the "iron triangle" dynamics we saw with Carlyle Group: government, industry, and influencers in a tight knot. Sobel points out that the Whiz Kids' data-driven ethos, while efficient, often ignored human factors, like troop morale. *When McKinsey Comes to Town* goes further, exposing conflicts of interest (McKinsey advised Chinese firms while working on U.S. military projects, raising eyebrows in Congress). By 2025, firms like Booz Allen Hamilton, a defense consulting heavyweight, were raking in $6.8 billion annually from DOD contracts. It's a system where consultants became the glue, linking primes like Lockheed Martin to Pentagon brass, but at the cost of billions in taxpayer dollars for advice that often presents as extremely busy PowerPoint slides.

The new Trump Administration and Secretary of Defense Pete Hegseth have declared war on "nonessential" consulting contracts, echoing a broader push to gut government spending. A February 2025 memo from the General Services Administration (GSA) targets top firms, demanding agencies justify $65 billion in contracts. A strange approach, though has been the Administration requesting that contracted companies suggest streamlining and canceling their own consulting contracts. If the consulting firms have been doing their legitimate best to satisfy the demands of the contracting agency, then the Government needs to identify which contracts it can live without. If the firms are profligate criminal waste, then why would anyone think they could reasonably excommunicate themselves?

Criticism of consulting providers is nothing new. Think tanks like CSIS have long flagged the DOD's consultant addiction. But Trump's team takes it to eleven, arguing firms like Accenture (which shed DEI policies to align with Trump's directives) are too entrenched. Early in this Administration, policies seem to be shifting daily. One constant truth is green suiter headcount can't reduce while also discarding DOD civilians and consultant staffing. The Trump administration's ire reflects a broader reckoning. Consulting's roots in the DOD, from the Whiz Kids to McKinsey's millions, built a system where outside expertise is both savior and crutch. Firms shaped platforms to last, but their influence, lucrative and sometimes murky, has sparked a backlash. By 2025, the Pentagon's at a crossroads: keep leaning on Bain's brainiacs or rebuild its own muscle. Either way, the legacy of consulting's grip on defense isn't fading quietly.

◈ ◈ ◈

## Contract Types

If there is an exciting way to describe contract types I have not found it yet. Contracts are the legal glue binding the DOD to large prime like Lockheed Martin or small machine shops like Whelan Machine. Cibinic and Nash's *Formation of Government Contracts* does not make a good coffee table discussion piece, but it is one of the primary texts used in acquisition programs at the University of Virginia, George Mason, and many others. The source document binding on the Government is the Federal Acquisition Regulation. First, the main contract types used in defense are Firm-Fixed-Price (FFP), Cost-Reimbursement (abbreviated to CP for Cost Plus), Indefinite Delivery/Indefinite Quantity (IDIQ), and Time and Materials (T&M). The growing use of Other Transaction Authority is worthy of its own section later. The DOD has recently described leaning harder on FFP and contractors are not enthusiastic about absorbing that risk.

FFP is the Pentagon's go-to when the job's clear-cut, like buying standard ammo or building a batch of radios with set specifications. The contractor agrees to deliver at a set price, no matter the hiccups or global pandemic. Think of it like booking an Uber at a locked-in fare, then when your driver gets stuck in traffic, she must absorb the lost time at no additional cost to the rider. FAR Part 16.2 says FFP puts "maximum risk" on contractors, forcing them to control costs to pocket a profit. It's low-maintenance for the DOD, with minimal oversight, but brutal for firms if supply chains snag or labor costs spike. Cibinic and Nash note FFP's popularity for commercial items, where specs are stable, but warn it's dicey for complex projects with murky requirements.

Cost-Reimbursement contracts, or CP, are the opposite. They are more like a blank check with guardrails. They are intended for R&D-heavy programs, like developing hypersonic missiles. CP lets contractors bill actual costs plus a fee, but unlike the commercial SaaS world this fee percentage is usually capped under 10%. FAR Part 16.3 lists types like Cost-Plus-Fixed-Fee (CPFF), where the fee stays static, or Cost-Plus-Incentive-Fee (CPIF), which dangles extra cash for beating cost targets. The DOD accepts CP for cutting-edge tech because it allegedly shares the risk burden with the contractor. But it's a headache to manage, needing audits to prevent padding. Also, there is no incentive for the contractor to ever finish; they can present add-on features and promote scope creep to keep the program churning for many years. Winning a program is a major investment so why would any company want to rush to finish and turn off their steady revenue stream?

Then there are Indefinite Delivery/Indefinite Quantity (IDIQ) contracts, the Swiss Army knife of defense contracting. IDIQs set up a framework for future orders without locking in quantities, perfect for ongoing needs like IT services or spare parts that may have wildly fluctuating demand needs over time. FAR Part 16.5 explains IDIQs can be single-award or multiple-award, with contractors competing for task orders. A 2018 GAO report says 40% of DOD obligations from 2015 to 2017 went to IDIQs. They're flexible, but Cibinic and Nash caution that vague terms can spark disputes over pricing or scope.

Time and Materials (T&M) contracts are the wildcard, used when the job's scope is fuzzy, like emergency repairs on a Navy destroyer. FAR Part 16.6 defines T&M as paying hourly labor rates plus material costs, with a ceiling to cap spending. It's high risk for the DOD, contractors could drag their feet to bill more hours, so it's a last resort. *Formation of Government Contracts* warns that T&M requires tight oversight to avoid abuse, and FFP's predictable budget spread is much easier to forecast in the POM. Think of T&M as hiring a plumber by the hour—you're on the hook if they take forever.

Since the 2017 National Defense Authorization Act (NDAA), the DOD's been all-in on FFP to curb cost overruns that plagued CP contracts, like the Littoral Combat Ship's budget bloat. Section 829 of the NDAA mandates a preference for FFP, requiring head-of-agency approval for CP contracts over $25 million after October 2019. The Defense Federal Acquisition Regulation Supplement (DFARS) backs this, citing a 2016 memo urging FFP for stable requirements, like foreign military sales. FFP keeps costs predictable for upgrades, unlike CP's open wallet.

But contractors aren't thrilled about swallowing all the risk. FFP's "maximum risk" clause means firms like Boeing or Raytheon eat losses if costs soar (like rare earth supply or changing tariff structures for raw materials). A 2019 Federal Register comment noted that contractors fear FFP stifles innovation, as they do not incentivize any contractor development above the bare minimum at the time of award. Cibinic and Nash highlight contractors' pushback: many demand economic price adjustments (EPA) in FFP deals to hedge against inflation or supply chain shocks.

This tug-of-war echoes the DOD's consulting battles with McKinsey, everyone wants predictability, but someone's gotta take the hit. Contractors, like BAE in York, lean on lobbying to soften FFP mandates. By 2025, the DOD's FFP crusade is reshaping defense. Boeing is one of the largest primes openly declaring it will not bid FFP development efforts.

❖ ❖ ❖

## The New Normal

Defense contracting is one of those industries where everything old becomes new again. A different contract type called Other Transaction Authority (OTA) and its newer cousin, Commercial Solutions Openings (CSO), are the supposed "fast lane" to getting the newest technologies from non-traditional defense contractors. Born to bridge the gap between the Pentagon and Silicon Valley, OTA and CSO are shaking up defense acquisition.

OTA started in 1958 when NASA's Space Act gave it a tool to team up with commercial firms for moon-landing dreams, free from FAR's shackles. The DOD hopped on in 1989 for research, with prototype powers added in 1993. The intent was to lure non-traditional contractors (think SpaceX or Google) that have a market cap and R&D budget bigger than the Pentagon's RDTE budget. Congress turbocharged OTA with the 2015 and 2016 National Defense Authorization Acts, making prototype authority permanent and allowing follow-on production contracts. CSO, introduced in 2016, is OTA's sleeker sibling, a streamlined process under 10 U.S.C. § 2371b for buying commercial solutions like software or drones. It's like a pop-up shop for defense tech, inviting pitches from anyone with a bright idea.

The DOD's OTA use has skyrocketed. From 3% of the R&D budget in 2015 to 18% by 2018. CSOs, managed by the Defense Innovation Unit (DIU), are catching up, with $1.2 billion awarded from 2016-2022 for projects like the Navy's Expeditionary Medical Platform. The OTA model is ideally suited for prototypes; 63% of funds from 2015-2019 went to mid-to-late-stage R&D. CSOs, meanwhile, grab off-the-shelf tech, like Palantir's data analytics, slashing delivery times. The DIU's CSO for a low-altitude communication system, awarded in months, would've taken years under FAR.

The OTA consortium model is a key player. Consortia are like innovation clubs—hundreds of companies, from Lockheed Martin to 3-person hobby shops, grouped under a manager like SOSSEC or ATI. The DOD signs a base OTA, then issues a call for white papers, picking winners for funding. The SOSSEC Consortium's COBRA OTA develops cyber defenses, while the Defense Industrial Base Consortium (DIBC) landed a $400 million propulsion deal. From 2019-2021, top consortia pulled in $15.6 billion, with the consortium collecting their few percent of pass-through fees.

For the DOD, OTA and CSO are a lifeline. They're fast projects like the Army's robotic mule prototype wrapped in a year, not a decade. They draw non-traditional firms, with 60% of DIU's CSO awards going to "newcomers." Flexible terms let the DOD negotiate IP or milestones, dodging FAR's one-size-fits-all rules. Follow-on OTAs, per the 2016 NDAA, streamline a transition to production. Contractors love the freedom, no FAR cost-accounting standards, and less paperwork.

But there's a catch. The DOD struggles with oversight. A 2022 GAO audit slammed poor tracking of OTA obligations, leaving billions unmonitored. CSOs, with their broad solicitations, risk vague requirements, sparking disputes in areas that don't have mature case law supporting protests or suits to the US Court of Federal Claims. Contractors face uncertainty: OTA's loose structure means sparse award feedback, and there's no performance rating system to support future bids. Then, the fact that losers on OTA bids cannot FOIA any information from the decision. Consortium fees, 3-5% for some, eat profits, and traditional firms grumble about having to bring substantial investment as their ticket to even enter and bid an OTA compared with a startup that just shows up with slides and Patagonia vests.

This OTA-CSO surge ties to our defense industry arc: rare earth materials, consulting cuts, and diminishing manufacturing sources all scream urgency for innovation. OTA flexibility, like the FFP push, seems to be focused on pathways around our own Soviet-style directed acquisition policy, and is not focused on the most pressing needs like defending Taiwan or Guam from growing PRC belligerence.

❖ ❖ ❖

## Small Business and Snake Oil

Ever wonder how a tiny startup could land a Pentagon contract without drowning in paperwork? That's where the Small Business Innovation Research (SBIR) program comes in, a government lifeline tossing non-dilutive funding to small businesses with big ideas. Born in 1982, SBIR is a brainchild of the Small Business Innovation Development Act, designed to spark innovation, boost small firms, and meet federal R&D needs. There are some promising incentives but it is a rocky road from Phase I to a program of record, across the Valley of Death, and fighting for scraps against "SBIR mills."

SBIR roots trace to a time when big defense contractor primes had a tight grip on almost all DOD R&D and small businesses struggled for a seat at the table. Congress, nudged by the Small Business Administration (SBA), launched SBIR to level the playing field. They mandated 11 federal agencies with R&D budgets over $100 million (like the DOD, NASA, and NIH) to set aside a slice for small firms. The intent was to fuel technological breakthroughs, encourage minority and women-owned businesses, and turn lab ideas into market-ready products. The DOD is the program's heavyweight with $1 billion in annual grants.

SBIR's three-phase structure is its backbone. Phase I, the proof-of-concept stage, dishes out up to $314,363 for 6-12 months to test an idea's feasibility. Phase II, with up to $2.1 million over two years, builds prototypes. Phase III is the holy grail: commercialization or federal adoption, funded by private investment or non-SBIR government contracts. The dream is taking a Phase I spark to a program of record, where technology becomes a Pentagon staple with its own Program Element in the budget. Unfortunately, that vision is a dream with almost no proof of ever occurring. The brutal gap between Phase II and III is called the Valley of Death. A 2007 National Research Council report notes only 10% of SBIR projects hit $5 million in sales, with most stalling at $1 million or less. One of the primary reasons is that the DOD buys capabilities at a grand scale, and those platform providers are large companies. They have no incentive to dilute their work share of a program to cut in a small startup with a lighter widget.

This brings us to "SBIR mills," a dirty little secret in the program. Some firms game the system, churning out Phase I and II proposals to snag grants without aiming for Phase III. These mills, often small outfits with savvy grant writers, focus on low-risk, incremental research. They produce slick reports rather than market-ready tech, and now with AI proposal writing, they can build off their existing data library for endless permutations of the same technology concepts. A 2018 GAO report hints at this, noting duplicate funding cases where companies "recycled" ideas across agencies, pocketing millions for redundant work. The DOD's partly to blame, favoring paper deliverables over hardware. A 2020 CSIS study gripes that SBIRs are heavy on "report-driven" outcomes, with only 30% of DOD awards from 2015-2019 funding physical prototypes. This skews toward software or studies, not the nuts-and-bolts hardware, like solid rocket motors, that the Pentagon needs to counter China's industrial growth.

The DOD is waking up to this. The 2024 National Defense Industrial Strategy pushes for more hardware-focused SBIRs with $1.4 billion allocated to spur "tangible tech." But mills and paper-heavy grants persist, siphoning funds from firms that actually build stuff. For small contractors SBIRs are a double-edged sword. Phase I and II grants are low-risk cash, but the Valley of Death looms large. Venture and Private Equity are catching on that a successful user demonstration to a Team of SEALs after Phase II does not turn into a program of record.

# CHAPTER 7

# Arsenal of Democracy

*"A billion here, a billion there, and pretty soon you're talking real money."*
—Senator Everett McKinley Dirksen

THE ARSENAL OF DEMOCRACY, A term coined by President Franklin D. Roosevelt in 1940, captures the United States' transformation into a global powerhouse of military production, a legacy that informed acquisition professionals and political amateurs both wish could be a reality again. This chapter explores the backbone of that capability: the defense industrial base (DIB). Roosevelt's term tried to rally America's industrial might for World War II. Today, the military-industrial complex is far more fragile, arguably losing ground to peer threats due to maintaining an industrial approach that does not align with the new reality.

The National Technical Industrial Base (NTIB), defined by 10 U.S.C. § 2500, is the backbone of specific defense-only production and adjacent enabling industries. There are the polished labs and cubicle farms of large primes, but also the gritty work of welding at arsenals like Watervliet, forging cannon barrels. These facilities, often shielded by congressional champs, have battled several rounds of Base

Realignment and Closure. Now, the Trump administration's Department of Government Efficiency may be finding a new path to BRAC by another name, with GSA posting government building auction links to social media.

Shipyards, from Newport News to Bath Iron Works, are another linchpin, building Virginia-class subs and Zumwalt destroyers. These under-resourced shipyards were a shining example of private industry literally building power projection but now can't keep up with even anemic demand. Access to raw materials, even cotton that our artillery propellant requires comes from a specific type of Canadian tree. Then, the Defense Production Act is like the tape on infomercials slapped onto large jugs of water to stop springing leaks. It definitely works for a short time, but building pressure will always find new ways to overflow the fragile defense industrial ecosystem.

※ ※ ※

## Depots and Arsenals

The origins of U.S. arsenals date to the Revolutionary War, when a fledgling nation needed arms to fight off the British. That was when America decided tea should only be served two ways: sweet or at the bottom of Boston Harbor. The Springfield Armory, established in 1777 in Massachusetts, led the charge. It was selected by George Washington for its water-powered mills and strategic perch. Watervliet Arsenal became the cannon capital, forging barrels for the War of 1812 through to today. Shipyards, like the Brooklyn Navy Yard (1801), built frigates to fend off British fleets.

During the Civil War, arsenals like Rock Island and depots like Anniston were cranking out rifles and storing ammo while shipyards hammered out ironclads like the USS Monitor. World War I saw arsenals scale up, with Watervliet producing artillery for the trenches. But it was World War II that supercharged the system. Herman Moore's *Freedom's Forge* and *A Call to Arms* by Maury Klein are the seminal sources for describing the American industrial mobilization for World War II. They detail how depots like Tobyhanna became logistics nerve centers, repairing radios and storing tanks, while shipyards like Kaiser's Richmond yards pumped out 747 Liberty and Victory ships.

Enter Rosie the Riveter, the bandana-clad icon of WWII's home front. As men marched to war, six million women flooded factories and shipyards, riveting B-24 bombers and welding destroyers. At Ford's Willow Run facility, bomber aircraft rolled almost continuously off the line. Shipyards like Mare Island set records, launching the USS Ward in 17.5 days.

Post-WWII, arsenals and depots adapted. The Korean and Vietnam Wars leaned on depots like Red River to overhaul helicopters, while shipyards built carriers like the USS Enterprise. Jacques Gansler's *The Defense Industry* highlights their Cold War role, producing missiles and electronics as threats shifted from tanks to nukes. Today, Watervliet forges M1 Abrams barrels, and Norfolk Naval Shipyard refits nuclear subs.

In the modern era, there are generally two flavors of arsenals and depots. The first is GOGO: Government-Owned, Government-Operated. This means the DOD runs the show, like Watervliet where federal employees operate the machines. The other model is GOCO: Government-Owned, Contractor-Operated. This means that large contractors like General Dynamics bid on 10-year contracts to provide the staff and maintain the facilities for the DOD. There have been eras when each management model was arguably a success, and several where they failed. Former Under Secretary of Defense for Acquisition, Jacques Gansler credited the GOCO model back in 1980. The DOD's 2024 Industrial Base Strategy cites GOCO's success, with 80% of ammo plants contractor-run. An area where this model crumbles is determining who bears the modernization costs for crumbling infrastructure. The GOCO operator is like a renter in an apartment, they may paint and put down some rugs but they don't want to put on a new roof. The Radford Army Ammunition Plant is a perfect example of GOCO failing to keep pace with demand. The facility still uses wooden buildings from WWII and hand-pushed carts to move explosive materials around.

Klein pointed to Watervliet's bloated costs back in the 1970s, with unionized workers slowing modernization. GOCO's agility comes with risks, the daily work orders submitted by the operator must be approved by on-site government representatives. Gansler warned that GOCO's reliance on single contractors, like at Holston Army Ammunition Plant, create vulnerabilities as contractors adjust their financial decisions and strategies over the course of long periods of performance.

This ties to our broader call to action: arsenals and depots are elements in an enterprise desperately trying to find strategies to counter growing peer threats. Too much government control, and you get sluggishness; too much contractor freedom, and you risk corner-cutting. By 2025, the Arsenal of Democracy's depots and shipyards remain vital, but they're wrestling with an aging base, political pressures, and a world where industrial might is as critical as ever.

◆ ◆ ◆

# BRAC

Imagine a sprawling military empire, with bases dotting the globe, some humming with purpose, others sitting idle. That's the puzzle the Base Realignment and Closure (BRAC) process was designed to solve. BRAC was a congressionally mandated effort to trim the Department of Defense's fat, boost efficiency, and redirect billions to actual warfighting. Kicked off in 1988, BRAC reshaped America's defense industrial landscape, closing over 350 installations across five rounds. With the first in 1988 and the last in 2005. There were interesting locations targeted, and more than one dubious save. Congress and Senators played both sides, clamoring for closures in other areas but of course heavily pushing to keep their own. The Trump administration's DOGE has become a verb with how they are shedding civilian workforce and forcing the General Services Administration to shake up DOD real estate in 2025. It's a tale of politics, strategy, and the tug-of-war between national security and local pride.

BRAC's mission was to streamline the DOD's massive port-folio after the Cold War, when bases built for Soviet threats became costly dinosaurs. The process, authorized by Public

Law 100-526, created independent commissions to recommend closures and realignments, shielding decisions from political meddling. Targeted locations spanned iconic arsenals, shipyards, and airfields. The 1988 round hit smaller bases like Fort Sheridan, Illinois, while 1991 closed major sites like Philadelphia Naval Shipyard. The 1993 round axed Naval Air Station Alameda, California, and the 1995 round shuttered Bergstrom Air Force Base, Texas. The 2005 round, the biggest and costliest, closed 22 major bases and realigned 33 others, affecting 120,000 personnel. From Brooklyn Navy Yard in 1964 to Fort Monmouth, New Jersey, in 2005, BRAC supposedly targeted excess capacity.

The most significant closure was arguably the Brooklyn Navy Yard, shuttered in 1966. Though pre-BRAC, it set the tone for later rounds. Klein's *A Call to Arms* notes its WWII peak, building battleships like the USS Iowa, but post-war decline made it a budget drain sitting on very valuable real estate. Its 300-acre footprint, employing 70,000 at its height, was a loss felt across New York. The Navy Yard's closure, announced by Defense Secretary Robert McNamara in 1964, was a gut punch to Brooklyn. At that time, it employed 9,100 workers, and it was New York's oldest industrial plant, having built iconic ships like the USS Arizona and USS Missouri. When the gates closed on June 25, 1966, the City of New York stepped up, buying the yard from the federal government in 1969 for $24 million after a tug-of-war with the Johnson administration, which initially balked at the sale. The city's goal was to turn the yard into an industrial park to replace lost jobs.

The city tapped the Commerce, Labor, and Industry in the County of Kings (CLICK), a nonprofit, to manage the yard's rebirth. CLICK leased space to tenants like Seatrain Shipbuilding, which employed 3,250 workers to build supertankers in the 1970s. But Seatrain's collapse in 1979, amid strikes and mismanagement, left the yard floundering. Brooklyn House Representatives Shirley Chisholm and Fred Richmond fought to secure $40 million in congressional loans to save jobs, but the yard's early years were rocky, with employment dipping below 3,000 by 1980.

In 1981, Mayor Ed Koch replaced CLICK with the Brooklyn Navy Yard Development Corporation (BNYDC), a nonprofit that became the yard's enduring steward. By 2025, BNYDC manages a 300-acre campus with over 550 businesses, 13,000 jobs, and $2.5 billion in annual economic impact. While redevelopment may be commendable, that supposed excess shipbuilding capacity has been plaguing the United States in commercial and military shipbuilding.

The most questionable save that survived BRAC was Ellsworth Air Force Base, South Dakota, spared in 2005. Home to B-1B Lancers, Ellsworth faced scrutiny for its remote location and high operating costs ($1.2 billion annually). Critics, including CSIS, argued nearby bases like Minot AFB could absorb its mission, saving $200 million yearly. But South Dakota's congressional delegation, led by Sen. John Thune, pulled out all the stops. Thune, a rising GOP star, rallied local leaders and lobbied the BRAC Commission, touting Ellsworth's B-1 role and economic impact of 5,000 jobs. The save raised eyebrows; a 2006 GAO report flagged "inconsistent criteria" in sparing Ellsworth while closing similar bases. This base save shows the competing priorities of budget hawks and then the always important jobs numbers in Member home districts.

Congressional heavyweights shaped BRAC's outcomes, often blurring national and local interests. In 1995, Sen. Trent Lott fought to save Keesler AFB in Mississippi, leveraging his Senate clout to highlight its hurricane-tracking mission. Rep. Norm Dicks battled for Puget Sound Naval Shipyard, saving 8,000 jobs in his district. Congressional muscle often trumps strategic logic. Sen. Kay Bailey Hutchison pushed to realign Kelly AFB, Texas, in 1995. Gansler called this job redirecting to Lackland AFB a "political compromise." BRAC's independent commission was meant to curb this horse-trading, but lawmakers still found ways to flex.

Environmental cleanup and restoration, a BRAC staple, ballooned costs and evaporated the vast majority of so-called savings. The 2005 round, meant to save $35 billion, overspent by $22.8 billion due to underestimated restoration costs.

Fast-forward to 2025, and the Trump administration's DOGE is rattling the DOD's real estate cage. Paired with the GSA, DOGE is targeting $2 trillion in federal cuts, with DOD's $370 billion installation maintenance backlog on the watch list. A February 2025 GSA memo orders agencies to slash underused properties, with the DOD reporting 22% excess. DOGE is pushing to sell off bases like Fort Belvoir, Virginia, for commercial use. One thing is common across all eras in the defense industrial base. We have never had the exact right amount of capacity in the right locations.

❖ ❖ ❖

# Shipyards

*"Each of a hundred ships, built by the same men at the same yard to the same plans, will have her own special characteristics--most of them bad, really, but after her crew becomes accustomed to them they are spoken of affectionately, particularly in retrospect."*    —Alfred Thayer Mahan

From the early colonial days in the United States, we have been a coastal empire. Still now, the majority of Americans live within a few hundred miles of each coast. The ability to build ships and project naval power has been the pacing item for empires for thousands of years. Shipyards are the gritty sprawling hubs of steel and sweat where the U.S. Navy's might and the commercial shipping industry's backbone are forged. The defense industry is the lifeblood of these yards, pumping in billions to build warships and generate some level of national maritime capacity. US shipyard capabilities changed dramatically from the WWII heroics of Kaiser and Higgins, game-changing laws like the Two Ocean Navy Act shaped our strategic capabilities, and Congress is pushing to counter China's naval ascension.

The defense industry doesn't just support U.S. shipyards, it practically keeps them alive. Picture this: in 2019, shipbuilding and repair employed over 107,000 direct workers, generating $9.9 billion in wages and $12.2 billion in GDP. A whopping 95% of that cash flow comes from defense contracts, mostly for the Navy. Private yards like Huntington

Ingalls' Newport News are the heavy hitters, building everything from nuclear-powered carriers to submarines. But since the 1960s, 14 defense-related yards have shuttered, leaving just seven major private ones to carry the load. The Navy's orders keep these yards humming, but when contracts dry up, it's like pulling the plug on a city's power grid. Workers scatter, suppliers falter, and restarting is a nightmare.

The Pentagon is trying to plug these gaps, tossing $2 billion at suppliers and workforce training. The real challenge is that America's commercial shipbuilding is a ghost of its former self, producing just 0.2% of the world's ships. Compare that to China, which churns out 50% of global tonnage. High labor costs and aging facilities don't help. The ship count for the US Navy is widely talked about, whether it should be 300, 350, or 500 capital assets. Commercial freight-carrying ships and arctic ice cutters are also absolute driving lines in national power that never get the appropriate level of attention.

Let's step back to the 1940s when the world was on fire and America's shipyards were its arsenal. Henry J. Kaiser, a guy with zero shipbuilding chops, turned the industry on its head. He set up seven West Coast yards, four in Richmond, California, and three in the Pacific Northwest. Starting in 1940, Kaiser teamed up with Todd Shipyards and Bath Iron Works to build 60 cargo ships for Britain. Then came the Liberty ships, which were unglamorous but vital workhorses of the war. Kaiser's yards used assembly-line tricks, welding steel plates like a giant Lego set. The result was shipbuilding went from many months to only weeks. The SS Robert E. Peary holds the record, being assembled in four days flat. Kaiser's Richmond yards alone pumped out 747 ships and his total production was 1,490 vessels. His workforce investment in other areas like Kaiser Permanente healthcare still exists, but unfortunately, the ship capacity died over the past 50 years.

Then there's Andrew Higgins, the New Orleans maverick who gave us the Higgins Boat. These landing craft were the workhorses of the D-Day landing at Normandy. They had eight plants churning out 20,000 boats by 1945. His plywood designs were cheap, fast, and tough, perfect for storming Normandy or Iwo Jima. Eisenhower called Higgins' boat war-winners, and he wasn't wrong.

Our current shipyard footprint is scattered across America's coasts like strategic chess pieces. The Maritime Administration (an agency under the Department of Transportation) counts 21 major yards capable of building big ships: five on the Atlantic (think Newport News, Virginia, or Bath Iron Works, Maine), seven on the Gulf (like Ingalls in Mississippi or Austal USA in Alabama), five on the Pacific (NASSCO in California), and four on the Great Lakes (Fincantieri Bay Shipbuilding in Wisconsin). The "Big Six" are Newport News, Ingalls, Bath, Electric Boat in Connecticut, NASSCO, and Avondale in Louisiana. These yards handle 90% of military work and receive two-thirds of the industry's order volume. The South Korean conglomerate Hanwha has made its move into the US ship industry by purchasing the Philly Shipyard.

Shipyards aren't just factories, they are truly the foundation of our entire concept of military power. We take it for granted the American ships have complete freedom of navigation around the globe. The US Navy has under 300 deployable ships, but their plans and supporters in Congress are trying to address a shrinking Navy as ships are retired. The Congressional Budget Office says it will be $40 billion a year for the Navy to reach its ship count objective.

For commercial shipping, shipyards keep the Merchant Marine alive, hauling goods and supporting wartime logistics. The Jones Act, passed in 1920, mandates U.S.-built ships for domestic trade, protecting yards but raising costs. American ships are 2–3 times pricier than foreign ones. The biggest reason for the difference is China's subsidization of their ship industry. With US commercial shipping output at 0.2% globally, we're leaning hard on foreign shipping, which is risky when the Administration is also trying to completely rescope the global trade policy. The White House's 2025 Maritime Action Plan wants to change that, pushing for more commercial builds to bolster the economy and back the Navy.

The Two Ocean Navy Act of 1940 was a dramatic turning point. With Japan and Germany flexing their muscles, Congress greenlit a 70% boost in Navy tonnage. This funded 257 ships including 7 battleships, 18 carriers, and 115 destroyers. It wasn't just about ships; it poured money into yards, sparking a building frenzy. By 1945, nearly 6,000 vessels, including Liberty and Victory ships, were afloat. The Merchant Marine Act of 1936 subsidized 50 commercial ships a year for naval use, bumped to 200 by 1940. The Jones Act keeps domestic trade American-made, but it's a double-edged sword, pricing us out of global markets. Recent moves, like the expired Shipyard Act of 2021 and the proposed SHIPS for America Act, aim to modernize yards by incentivizing defense industry investment. The

2025 Executive Order on Maritime Dominance is pushing a 180-day plan to use the Defense Production Act and lure investors.

China's navy is growing like wildfire, and Congress is scrambling to respond. The FY2025 NDAA boosted shipbuilding budgets by 12.5% annually from 2020–2024, rejecting the Navy's Shipyard Accountability plan for lack of clarity but doubling down on workforce and supplier support. Programs like the Columbia-class submarine are lagging, and Congress wants answers. China's dual-use yards, blending commercial and military, give them an edge. Their 84 subsidiaries and 200,000 workers dwarf our national capacity.

Huntington Ingalls Industries (HII) is the king of U.S. shipbuilding. Its Newport News yard, founded in 1886, builds nuclear aircraft carriers and submarines. Ingalls, in Mississippi, cranks out destroyers and amphibious ships. HII's revenue in 2024 was $11.5 billion, and they have an additional $48.7 million in backlog work. Some argue the demand is only weak because there is not enough workforce or capacity to even approach additional shipbuilding.

The South Korean giant Hanwha is jumping into the US market, buying the Philly Shipyard for $100 million in 2024. The Philly yard will allow them to build Jones Act-compliant ships and still rely on their 200,000 global employee and technology base. Foreign ownership and investment in the defense industry is difficult. Several laws ensure that foreign defense subsidiaries in the US have some level of separation. A Special Security Agreement is used when a U.S. company with classified contracts faces Foreign Ownership, Control, or Influence (FOCI). It's like a firewall, ensuring foreign stakeholders can't access sensitive info or sway operations. The Defense Counterintelligence and Security Agency (DCSA) enforces SSAs, requiring U.S. citizens board majorities, cleared key managers, and strict oversight to block foreign meddling. Think of it as letting a foreign owner hold stock but not the reins.

FOCI, meanwhile, is when a foreign entity could influence a U.S. company's decisions, risking classified data or contract performance. It's not just majority ownership; even minority stakes or foreign loans can trigger FOCI concerns. The DCSA evaluates risks, like espionage or tech transfer, and tailors mitigation, from board resolutions to SSAs, to keep the company secure. This is a major administrative hurdle that Hanwha is taking on. BAE Systems and Rheinmetall seem to have cracked the code. Singapore's ST Engineering has a North American subsidiary structure, but their close ties to mainland China will prevent them from becoming a true US defense player.

Shipyards are America's maritime muscle, powered by the defense industry and shaped by legends like Kaiser and Higgins. From Newport News to Philly, these yards keep the Navy strong and trade flowing, but they're up against aging infrastructure and China's relentless growth. Laws like the Two Ocean Navy Act set the stage, and Congress is pushing hard, but the gap with China's navy is real. With HII leading and Hanwha shaking things up, the future's uncertain. Can America's yards rise to the challenge? That's the question, and the answer's being welded together, one ship at a time.

❖ ❖ ❖

## Raw Materials and Access

The defense industry runs a constant global scavenger hunt for raw materials like rare earths and titanium, needed in the platforms and aircraft. The Pentagon's supply chain stretches across continents, navigating geopolitical minefields and trade wars. Right now, rare earth minerals are making headlines, thanks to a new Trump executive order. Silicon chips, powered by TSMC, connect our warfare dependence to a nation directly in China's growth plan. This isn't a new story; with a legacy of sourcing gun cotton from Canada or the CIA's sneaky titanium deals with the Soviets for the SR-71. But with tariffs shifting like quicksand, the defense industry's facing a tough road ahead.

The U.S. defense industry is like a master chef, sourcing exotic ingredients from every corner of the globe to whip up high-tech weaponry. Critical minerals like titanium, tungsten, cobalt, and rare earths are the spices that make it all work. The geographic reality is the U.S. doesn't have enough of these at home. China controls 70% of rare earth mining and 90% of processing, while places like the Democratic Republic of Congo dominate cobalt. In 2019, the defense sector consumed $1.5 billion worth of these materials, with 95% of rare earths imported, mostly from China. The Pentagon's been scrambling to secure supplies, friend-shoring with allies like Canada and Australia, but it's a slow grind.

Geopolitics made this difficult even before the new Administration's dramatic international trade actions. The lithium vital for batteries in drones and electric vehicles often comes from Argentina and Chile. China subsidizes this trade to seize every available export, while the US is now punishing those countries for not buying enough US exports. The Defense Logistics Agency (DLA) stockpiles 47 commodities, worth $1.5 billion, to hedge against disruptions. The industry is also investing in recycling and substitutions like aluminum for copper in some munitions. But it's like swapping sugar for stevia; it doesn't always taste right.

Lately, rare earth elements like dysprosium, neodymium, and yttrium are stealing the show, and not in a good way. These metals are critical for high-temperature magnets in jet engines and radar systems, but China's got a stranglehold, supplying 80% of U.S. needs. On March 20, 2025, President Trump signed an executive order invoking the Defense Production Act to boost domestic mining and processing, aiming to cut reliance on foreign adversaries. The order fast-tracks permits, funnels funds to projects like California's Mountain Pass mine, and eyes deals with Ukraine for its lithium and titanium. It's a bold move, spurred by China's response to Administration calls for 145% tariffs on Chinese goods (apparently US trade and tariff policy changes daily through posts to social media, so this statistic will surely be outdated by the time you read this). Disruptions to F-35s, submarines, and smart bombs are expected with China's trade responses. The Center for Strategic and International Studies warns over a dozen defense firms could take a hit. MP Materials, running Mountain Pass, stopped shipping ore to China due to 125% counter-tariffs, banking on expanding U.S. processing instead. But building mines takes years, maybe decades. Can the U.S. really break free from China's grip before the next crisis? The executive order's a start, but it's like trying to build a dam during a flood.

Silicon chips are the brains behind every modern weapon. In *Chip War* by Chris Miller, the semiconductor industry is painted as the backbone of global power. Chips run missile guidance, drones, and even the Pentagon's coffee machines. Taiwan's TSMC, the world's chip-making titan, produces 90% of advanced semiconductors, including those for U.S. military gear. Miller notes that a single TSMC fab can churn out 100,000 wafers a month, each packed with billions of transistors. But here's the problem: Taiwan's a geopolitical hotspot, and a Chinese invasion could cripple U.S. defense overnight.

The CHIPS and Science Act of 2022 was touted as a $280 billion lifeline to bring chip-making home. It's funding TSMC's $100 billion Arizona fab complex, set to produce cutting-edge chips by 2028. Miller argues this is critical, without domestic capacity, the U.S. is at China's mercy for critical materials like gallium. *Chip War* highlights how the U.S. lost its semiconductor edge by the '90s, outsourcing to Asia for cheap labor. TSMC's Arizona move is a step, but with only 10% of global chip capacity stateside, it's a long climb. The Pentagon's sweating, chips are as vital as bullets, and they're running low.

International defense industry dependencies aren't new. The Hyde Park Declaration in 1941 connected the Canadian and US economies to increase Canadian imports of lumber and other materials to the US. Then, materials were exported from the US for final assembly in Canada for inclusion in the Lend-Lease program supporting the UK.

The wildest story of foreign trade for military dominance is how the CIA sourced titanium from the Soviet Union for the SR-71 Blackbird. In the 1960s, titanium's strength and heat resistance made it perfect for the Mach 3 spy plane, but the U.S. had no reliable supply. So, the CIA set up shell companies, funneled cash through third countries, and bought titanium from the USSR. By 1965, 90% of the SR-71's titanium came from Soviet mines, unbeknownst to Moscow. The defense industry has always played a global game, bending the rules to get what it needs.

The defense industry is often a messy exception to many trade agreements. The industry tries to comply with things like the Buy American Act and Berry Amendment, but often requires exception based on the requirements from the DOD. Starting in 2025 the new Administration has embarked on a tariff showdown with allies and competitors. China is responding, slapping export controls on rare earths like gallium, germanium, and tungsten, critical for defense and tech. The New York Times called it a "warning shot" across the Pentagon's bow, with firms like Lockheed Martin and Raytheon facing supply chain chaos. The Center for Strategic and International Studies says tariffs could make materials too pricey, choking industries like battery manufacturing. Imagine trying to build a drone when the price of dysprosium spikes tenfold, like rare earths did in 2012.

The shifting tariff landscape is a nightmare for planning. Defense contractors need stable supply chains, but tariffs are like playing whack-a-mole: hit one problem, and another pops up. Friend-shoring with Canada or Ukraine could help, but for material like Germanium, but the Canadian government seems to be bristling at comments about becoming a state. They have responded with boycotts and counter-tariffs. Plus, tariffs on allies like South Korea, who use Chinese rare earths, could snarl things further. The CHIPS Act's chip push is at risk, too. TSMC's Arizona fabs need gallium, and there is not a strong domestic source. From rare earths to silicon chips, the U.S. is racing to secure supplies while dodging geopolitical punches. Trump's executive order and the CHIPS Act are bold plays, but they're up against China's dominance and a tariff strategy that's stirring the pot.

❖ ❖ ❖

## DPA

The U.S. has a very unique tool available that sounds more like a Soviet economic approach than free-market ideals. Picture a president with a magic wand, redirecting factories and resources to crank out tanks, jets, or even ventilators in a pinch. That's the Defense Production Act (DPA) in a nutshell. It is a law that has been the Pentagon's ace in the hole since the Korean War. It's not just about guns and ammo. Tools like DPAS ratings keep the defense industry constantly re-prioritizing.

Back in 1950, the Korean War was raging, and the U.S. needed to ramp up production fast. Think steel for tanks, rubber for tires, and food for troops, all in short supply. The US had started building up production for World War II back in 1939, so it had a running start. The Korean War did not have this same timeline. President Truman signed the Defense Production Act into law, modeled after WWII's War Powers Acts. It gave the government unilateral authority to prioritize contracts, allocate materials, and even expand factories. The idea was to make sure the military gets what it needs, no questions asked. It wasn't just about war; the DPA covered natural disasters and economic shocks too. By 1953, it helped churn out 35,000 aircraft and 20,000 tanks.

The DPA wasn't a one-hit wonder. It set up a framework with seven titles, three remain active today: prioritizing contracts (Title I), expanding production (Title III), and regulating prices (Title VII). It's like a Swiss Army knife for crises, flexible enough to tackle anything from pandemics to trade wars. The DPA has taken a few different forms over the last 70 years.

The DPA has been tweaked over 50 times since 1950, like a car getting new parts to stay roadworthy. Early on, amendments focused on Cold War needs like 1955's push for missile production. Fast-forward to 2018, and space tech was added, reflecting the Pentagon's eye on satellites and Starlink-style networks. Each change made the DPA broader, covering not just metals and munitions but semiconductors, rare earths, and even medical supplies. In 1992 it was expanded to include homeland security in the post-Cold War environment. As the response to 9/11 took shape, the DPA was changed again in 2003. Then, in 2020, the Pandemic response caused changes to the authorities and statutes, funneling funds for masks, vaccines, and the Warp Speed program.

A major impact on the defense industry is the Defense Priorities and Allocations System (DPAS). DPAS ratings prioritize orders with two codes: DO (critical) and DX (highest national urgency). A DX-rated order for F-35 parts jumps the queue, forcing companies like Lockheed Martin to deliver before anything else. DO ratings cover most defense needs and DX ratings throw off all delivery schedules. But these ratings absolutely have major contractual impacts on suppliers.

DPAS ensures the Pentagon gets its gear prioritized over commercial orders. Any knock-on effect to other order timelines are at the expense and risk of the vendor. Firms must comply within days, or face penalties. A small supplier making bolts for both Boeing and Walmart might have to halt commercial work for a DO-rated Navy contract, eating into profits. In 2019, DPAS handled 300,000 rated orders, showing its reach. For big players like Raytheon, it's business as usual; for smaller shops, it's a scramble.

Fast-forward to 2025, and the DPA is making industry headlines again. President Trump's March 20, 2025 executive order is a big move, invoking Title III to boost domestic production of rare earths, semiconductors, and critical minerals. The order's funneling $2 billion to projects like MP Materials' Mountain Pass mine and TSMC's Arizona chip fabs.

The Pentagon is also using Title I to prioritize contracts for shipbuilding and hypersonic weapons. With China's navy at 370 ships and counting, the U.S. needs its yards like Newport News and Ingalls firing on all cylinders. In 2024, the DPA allocated $500 million to expand titanium and tungsten production, critical for missiles and armor. The White House's 2025 Maritime Action Plan leans on DPA authorities to modernize shipyards, offering loans and grants to firms like Huntington Ingalls. It's not just about weapons, the DPA is backing battery production for drones and electric vehicles, with $300 million for lithium projects in Nevada.

The DPA has been around for 75 years, yet it feels more relevant than ever. From Korean War tanks to COVID ventilators to today's chip race, it's a lifeline for national security. The important question will be how the current administration and the next view this governmental authority. The threats from China, Iran, Russia, and other emerging powers have access to asymmetric technologies and the scale of hardware the DPA was intended for will likely not determine the outcome of those competitions.

# CHAPTER 8

# Art of Selling to the Pentagon

## Paper Tigers, Golf Courses, and 10-Year Sales Cycles

THE DEFENSE INDUSTRY EMPLOYS A unique "selling" method to the Pentagon, blending relationship-building, technical expertise, and long-term strategy. Unlike consumer sales, it targets a single monopsony buyer across a 10-year cycle tied to the Acquisition Lifecycle and PPBE process. One of the major influences on the sales process to the DOD is the "generational" nature of major platforms. Ships, tanks, and aircraft are huge long-term investments, and the competitions to get a new platform is rare. In the modern era, the DOD never procures full-rate production quantities of platforms from more than one supplier. So if you are General Dynamics and the DOD plans a new tank to replace the Abrams, you have to use every tool available to defend your market share. For many reasons (logical and political) platforms have been maintained for 50-plus life spans. This means selling to the Pentagon is an extremely high stakes full-contact sport. Ask Grumman about their loss in the competition for the Advanced Tactical Fighter (that ended up becoming the F-22). That loss and the Last Supper events of the 1990s ended up with Grumman being acquired by Northrop.

Depending on who you ask, selling to the Pentagon is just like any other sales role ...or it is a completely different sport requiring unique, unicorn-type skills. All consumers have perceptions about sales from their everyday lives: buying a car, buying a house, shopping. In the professional world, there are all kinds of sales models: subscriptions, online ordering, and even door-to-door cookie dealers called Girl Scouts. While the "art" of selling to the DOD can be unique, much of the "science" matches other industries closely.

The defense market operates under a complex regulatory framework, with stringent compliance requirements and a highly structured procurement process. Government contractors must adhere to the Federal Acquisition Regulation, the Defense Federal Acquisition Regulation Supplement, ITAR export laws, and hundreds of other constraints. This is the start of the severe separation between the skills and requirements of a salesperson in defense and a salesperson in other industries.

One key difference between selling to the DOD and other industries is the sales cycle length. Defense procurement can take several years from the initial concept to the actual contract award, requiring a long-term, strategic approach to business development. Successful defense contractors must be adept at shaping requirements, building relationships, and positioning their offerings to align with the DOD's evolving needs and priorities. Suppose a major ground vehicle program takes ten years to design, build, test, and field. In that case, the competitors can be pretty sure there will be a few administrations (and their stakeholder appointees) changing throughout the program.

The starting point of most programs that get fielded to the US military begins with a business developer "selling the dream" of what *could* be. These individuals identify and qualify opportunities, shape customer requirements, and develop winning strategies to secure contracts. By leveraging their knowledge of the DOD's procurement process, customer relationships, and competitive landscape, these professionals help their companies navigate the complexities of the defense market, the dynamics of defense trade shows and demonstrations, and the importance of building relationships with key stakeholders. This chapter is not a "how to" guide to business development for the DOD. Luckily, that book was already written. *BD 360°: A New Spin on Federal Sales* by Amber Hart and Lisa Shea Mundt really does an excellent job of simplifying the terms and aspects of finding and competing for DOD opportunities.

On the other end of the spectrum is the cautionary tale *War Dogs*. It serves as a humorous reminder of the importance of ethical selling to the Pentagon. Based on actual events, the film follows the journey of David Packouz and Efraim Diveroli, who won a $300 million contract from the Pentagon to arm host-nation forces overseas during the Global War on Terror.

❖ ❖ ❖

## Definitions

First, we need to define a few sales terms that can have significantly different meanings in other industries. The first is Business Development (BD). Business development in the defense sector is a strategic function that involves identifying new business opportunities, understanding market trends, and positioning the company to win contracts. Describing the role as strategic is dependent on the company and types of offerings. Many companies in the field will not have a single employee with the word "Sales" in their job description. Business development professionals are often the front-line representatives of their companies, responsible for building and maintaining relationships with potential government customers, partners, and other industry stakeholders. They work to understand the needs and priorities of the military branches, aligning their company's capabilities with these requirements.

The role of a business development professional in defense typically involves market research, attending industry events and conferences, participating in government-industry days, and maintaining a deep understanding of the defense acquisition process. They work closely with technical experts within their organization to develop solutions that meet emerging government needs. Business development teams are also responsible for tracking upcoming opportunities, analyzing requests for information (RFIs), and contributing to go/no-go decisions on potential bids.

On the other hand, capture management is a more focused and intensive process that begins once a specific opportunity has been identified. Capture managers lead the effort to position the company to win a contract. This process often starts long before a formal request for proposal (RFP) is issued, sometimes years in advance for major defense programs. Capture managers work to develop a deep understanding of the customer's needs, preferences, and evaluation criteria. They orchestrate the development of a winning strategy, which includes identifying the right team members, potential partners or subcontractors, and critical discriminators that will set their solution apart from competitors. Capture managers often engage in customer meetings, competitive analysis, solution development, and pricing strategy.

As the opportunity progresses, capture managers transition into proposal development, working closely with proposal teams to ensure that the bid aligns with the capture strategy and effectively addresses all customer requirements. They play a crucial role in reviewing and shaping the final proposal to maximize the chances of winning the contract. Depending on the company's size, BD, Capture, and Proposal Manager could be the same Army of One individual. The large primes are mainly divided into distinct departments and career tracks.

Both business development and capture management roles underscore the importance of building and maintaining solid relationships with government customers. In the defense industry, these relationships are built on trust, credibility, and a demonstrated understanding of the customer's mission and challenges. A unique aspect is that there is no single buyer with decision authority for large defense procurements. The bigger the program cost, the more layers and participants on the customer side. For example, the program office for the F-35 Joint Strike Fighter allegedly has more personnel in the Pentagon than the actual Air Force Headquarters.

Marketing in the defense industry differs significantly from business-to-consumer marketing. While consumer marketing often focuses on emotional appeals, brand awareness, and reaching a broad audience, defense marketing is much more targeted and technical. The audience for defense marketing is typically a small group of government decision-makers and influencers rather than the typical buying personas used in other industries. This is why you don't see many Northrop Grumman ads in your Facebook newsfeed, but the advertisements on every wall of the DC metro for a few stops around the Pentagon are permanently booked out by major defense primes.

The combined effort of these signs up in the Metro, the booths at trade shows, and sponsoring golf outings is all part of chasing the "perfect sale." The Grand Slam that defense BD professionals are chasing would be a new program of record, procuring an IRAD-developed technology so they can keep all IP, receive commercial item justification for preferred pricing, an Urgent Operational Needs Statement calling for immediate fielding, and a J&A for sole source procurement.

❖ ❖ ❖

# Big Game Hunting with Shipley®

The defense industry has "War Rooms." They are usually in bland office buildings in Arlington, Virginia, and don't have any of the Dr. Strangelove command center vibes. The teams of strategists, engineers, and writers make up the "capture team." Deciding whether to bid or not bid on an opportunity is a gamble for the company where the stakes are millions and sometimes billions of dollars. The defense industry's business development process is a high-wire act, blending strategy, creativity, and hard analysis. One of the main templates and doctrines of this process comes from Shipley Associates. They have a structured, dynamic framework that guides companies from spotting an opportunity to sealing the deal. The process of capture team, bid and proposal funding, proposal writing (with a spotlight on Sections L and M), probability of win (PWin) estimation, and game theory make it a very sporty segment of the industry.

The Shipley BD Lifecycle is like a battle plan for defense contractors. It's split into phases: Market Identification, Opportunity Assessment, Capture Planning, Proposal Planning, Proposal Development, Post-Submission, and Contract Award. Each phase builds on the last, turning a flicker of opportunity into a signed contract. What makes it compelling is its balance of structure and flexibility. Companies like Lockheed Martin or Raytheon use Shipley (or their own homegrown spin-off version) to stay disciplined while adapting to the chaos of government RFPs that may only provide 30 days to respond and often get released on Friday afternoons before long weekends.

The capture team is the nerve center of the proposal process. Think of them as a SWAT team: a mix of program managers, technical experts, pricing gurus, and BD pros. Their job is to position the company to win before the RFP even drops. A typical team might include a capture manager (the quarterback), a solutions architect (the tech wizard), and a contracts specialist (the rulebook nerd). They're obsessive about understanding the customer's preferences, and if they have time they develop impressive statistics on previous contracts and vendors.

The capture process is part detective work, part chess. The team digs into the customer's priorities: Are they obsessed with cost? Innovation? Speed? They meet with agency officials, attend industry days, and scour online tools like Jane's or GovWin. This intel shapes the winning strategy, which might mean teaming with a small business for set-aside points or highlighting past performance to build trust.

Capture teams live on coffee and stress. When an RFP is released, the team will usually sequester to their company War Room for the entire 30-45 days involved in the proposal response period. If the opportunity is a "must-win" like defending one of their key franchise platforms, it means 18-hour days, 7 days a week, until the final proposal gets sent off.

Bidding on a defense contract isn't cheap. Developing a proposal can cost anywhere from $100,000 to millions, depending on the contract's size. This is where Bid and Proposal (B&P) funding comes in. Companies set aside B&P

budgets annually, treating it like R&D for sales. It's a gamble, you're spending big with no guarantee of winning. Smaller firms are definitely at a disadvantage in the gamesmanship of well-crafted proposals. The funding process starts early, often during Opportunity Assessment. Leadership reviews the opportunity's potential return on investment. A $500 million contract might easily justify a $2 million proposal effort, once all the consultants and outside reviews are added in. Once approved, B&P dollars cover everything: capture team salaries, market research, consultants, and the proposal itself. This alone is a huge moat that keeps many of the best commercial companies on the sidelines. This is a high-risk process that is almost impossible for an outsider to properly set up initially. Many large companies will take a swing at an opportunity because it sounds perfect for them, but they are passed over by a middle-tier player in their industry that is organized specifically for this process.

When the RFP lands, it's game time. The proposal is your pitch, a 100-to-500-page beast that must convince the government you're the best choice. The Shipley process breaks this into Proposal Planning and Development. Planning means dissecting the RFP, building a compliance matrix, and assigning writers. Development is where the magic happens: crafting a narrative that's clear, persuasive, and bulletproof.

Sections L and M are the RFP's soul. Section L (Instructions, Conditions, and Notices to Offerors) tells you *how* to write the proposal. It's the rulebook, things like font sizes, page limits, and submission deadlines, with countless proposals thrown out almost every time for absurdly unimportant reasons. Miss a detail, and your proposal's trashed. Section

M (Evaluation Factors for Award) tells you *what* the government values. It lists criteria like technical approach, past performance, and cost, ranked by importance. Nail Section M, and you're halfway to winning.

Writing for Section L is like following a recipe. You format the proposal exactly as instructed, with limits to font size and page margins. Section M is trickier, it's where you flex your storytelling. Say the RFP prioritizes "innovative solutions." Your team might highlight a cutting-edge radar system, weaving in data on its performance and testimonials from past clients. You're not just listing features; you're painting a picture of success.

Probability of Win (PWin) is the highly subjective math of BD. It's a number, like 60%, that predicts your chance of winning. Calculating it is half science, half gut. The science part involves scoring your proposal against Section M criteria. Tools like Shipley's PWin calculator weigh factors like past performance, technical strength, and price competitiveness. If you've got a proven track record and a killer solution, your PWin climbs.

The art comes from intuition and experience. Capture managers lean on experience: Have we won similar contracts? Is the customer cozy with a competitor? Sometimes, it can be based on something as small as whether that program manager spent a lot of time at a competitor's booth at AUSA. PWin isn't static; it evolves as you gather intel. A competitor's slip-up (say, a late delivery on another contract) might bump your PWin from 50% to 70%.

Competitive analysis is where game theory sneaks in. You're not just selling to the government, you're outmaneuvering rivals. Game theory helps you predict competitors' moves and counter them. Imagine a three-way race between Lockheed Martin, Boeing, and Northrup Grumman for an aircraft platform. A "Black Hat" analysis is conducted by offerors to try and determine what the most dangerous course of action each competitors could take.

One vendor may be planning to bid artificially low to buy their way in. Another firm may have done a Joint Venture with a new star Silicon Valley startup to demonstrate a new approach. All of the usual MBA management consultant tools like SWOT analysis or Porter's Five Forces get brought into the calculus. The reason this process is so involved and existential to the bidders is that many of these opportunities are generational. If you lose the aircraft platform now, it may easily be 20-30 years before the next replacement makes its way back into the start of the requirements process to start again.

The Shipley model has stayed at the top of the pile using classes, certifications, and updating their doctrine when the federal process changes. From capture teams burning the midnight oil to writers agonizing over commas, it's a process fueled by ambition and teamwork. Winning a defense contract isn't just about money; it can truly be life or death for a company. As trends like AI and hypersonics reshape the industry, the lifecycle evolves, but its core remains: know your customer, outsmart your rivals, and tell a damn good story.

❖ ❖ ❖

## Where the Game is Played

The "game" of defense sales is definitely travel ball. A major component of a defense company's "go to market" strategy revolves around participating in the major trade shows. Picture sprawling convention halls with vehicles and weapons. Sleek drones gleam under spotlights, missile mockups loom over polished booths, and suited execs weave through crowds, clutching coffee and pitch decks. This is the world of defense industry trade shows like AUSA in Washington, D.C., or Eurosatory in Paris. There are plenty of critics like Anthony Sampson's *Arms Bazaar* and William Hartung's *And Weapons for All*. Professionals in the industry would say these events are less about flashy displays and more about strategic connections, especially with defense acquisition decision-makers. They're also not so different from trade shows in other industries, like tech or automotive. Chevy Chase's *Deal of the Century* is a funny send-up of an exaggerated yet oddly familiar hustle.

These trade shows are massive. AUSA, hosted by the Association of the United States Army, draws over 40,000 attendees annually to the Walter E. Washington Convention Center. Eurosatory, outside Paris, sees 100,000 visitors eyeing tanks and missile launchers. Sampson describes such events as modern arms fairs, evolving from secretive Cold War deals to open marketplaces where nations and contractors haggle. Hartung notes their global reach with shows like IDEX in Abu Dhabi or DSEI in London attracting everyone from Pentagon brass to oil sheiks. The vibe is part tech expo, part diplomatic summit, and part family reunion for military folks.

Booths are the heart of the action. The large primes will have multi-story setups with digital projections, real hardware, and plush meeting rooms tucked behind doors. Booths range from several million dollar construction to folding tables from small charities and vendors. A critic could argue it is somewhat perverse to present weapon technologies with the same glitz as a car show. Companies and foreign intelligence agencies have their intelligence collectors out taking pictures and looking for small details. A typical booth might feature a 3D-printed drone, a VR demo of a tank's targeting system, and a barista station with copper cappuccino machines.

There is almost nothing ever "sold" at a trade show. Meeting the right people is everything. Defense acquisition decision-makers like DOD program managers or Pentagon budget officials are coveted booth visitors. The DOD provides guest lecturers to give generally high-level overviews of objectives. Companies are in a rush to fill up their visit calendars to their booths, inviting the key decision-makers they want to talk within an open setting. Government personnel involved in source selections and ongoing competitions are generally barred from discussing those contracts, much like the basic response in lawsuits "we do not comment on pending litigation."

The stakes are sky-high. A single meeting at AUSA can spark a $1 billion deal, like Raytheon's Patriot missile upgrades. But it's not all glamour. Industry pros slog through 18-hour days, their feet aching from convention floors, their voices hoarse from pitches. Hartung captures the grind: reps juggle jet lag, cultural nuances, and ethical tightropes, especially in nations with looser regulations.

Defense trade shows aren't unique. CES in Las Vegas, where tech giants unveil gadgets, mirrors AUSA's energy. Car shows like Detroit's Auto Show, with their concept cars and exec schmoozing, feel very similar. At CES, Apple courts retailers; at AUSA, General Dynamics courts the Army. Both have glitzy booths, VIP lounges, and a race to impress decision-makers. The main difference? Defense shows carry a heavier geopolitical weight, decisions and announcements actually feed into the appearance and assessments of national power.

AUSA 2024 showcased quantum computing alongside tanks, reflecting the industry's pivot. Yet the core remains: booths to impress, decision-makers to woo, and deals to chase. Whether it's AUSA or CES, the game's the same, only the toys change. In 2025, the industry is reacting to the DOGE impact on travel bans for government personnel. This limits the main "front door" to interacting with the DOD on program concepts.

❖ ❖ ❖

## Navy SEALs and Network Engineers

These "Sales Jobs" in defense are nothing like Nicholas Cage's character in *Lord of War*. They are former Navy SEALs, network engineers, Ivy League MBAs, and all kinds of other archetypes. The job is to chase billion-dollar contracts, jet across the country for meetings, and navigate a maze of ethics laws to stay squeaky clean. Welcome to the defense industry, where business development, strategy, and account managers live and die by their most recent win. These folks are road warriors, living out of suitcases while juggling complex roles. Then there's the "revolving door" with the Pentagon, controversial but not unique. Ethics laws keep most corruption at bay, but when scandals like Boeing's KC-46 fiasco hit, they explode into headlines.

These pros are rarely home. Defense industry work is a whirlwind of travel, airports, rental cars, and hotel loyalty points. BD managers fly to D.C. for Pentagon briefings or to Huntsville, Alabama, to tour a helicopter repair facility. Account Managers hit industry conferences like AUSA in Washington or Sea-Air-Space in Maryland. Even proposal managers, who seem desk-bound, travel for kickoff meetings or color team reviews.

A capture manager will land in San Diego for a Navy RFP briefing, spend two days in meetings, and then hop a red-eye to Boston for a subcontractor negotiation. Their suitcase is their office, and their laptop automatically connects to Marriott Bonvoy wifi. Sales in the defense industry is the same road warrior life as many other industries, but the difference is there is an actual national security implication. If a Lockheed capture team fails at adequately preparing for a proposal, the US military may actually buy a less capable platform. A software BD rep selling subscriptions to dog-walking apps doesn't have that added layer of real-world relevance. The DOD doesn't just email out contracts to companies in some kind of rotation, it is a constant chase.

The "revolving door" describes how people move between government jobs (like at the Pentagon) and private industry. A DOD official might join Lockheed Martin as a BD exec or a Boeing strategist might become a Pentagon Assistant Secretary. It's a common criticism aimed at the defense industry saying it breeds conflicts of interest. But it seems to be fairly common across all regulated industries.

In 2022, Boeing hired 85 former government officials, more than any other defense contractor. This flow isn't unique. In finance, regulators join banks; in tech, Google snaps up ex-government talent. Knowledge is power, and companies want insiders who know how the Pentagon ticks. Meanwhile, government agencies benefit from industry expertise. It's a two-way street.

The U.S. has a web of laws to prevent corruption in defense contracting. The FAR sets strict rules for procurement, ensuring fairness. The Ethics in Government Act requires financial disclosures from senior officials to spot conflicts. Post-employment restrictions make senior DOD officials "radioactive" for a few years before they can approach their previous organizations on behalf of industry. But there have been a few recent scandals that show that even a rigorous process can't prevent all kinds of bad actors.

In the early 2000s, the Boeing KC-767 tanker deal became a notorious defense scandal, blending ambition with corruption. The Air Force needed to replace aging KC-135 tankers, and Boeing's KC-767, a modified 767 jet, was pitched in a $20 billion lease deal. Unlike outright purchases, leasing raised red flags for waste. Senator John McCain slammed it as a Boeing bailout, noting it cost $5.6 billion more than buying the planes.

The scandal broke in 2003 when Darleen Druyun, an Air Force procurement official, was caught favoring Boeing. While negotiating the deal, she leaked Airbus's bid details and inflated prices, later landing a $250,000 Boeing job. Druyun pled guilty, serving nine months in prison. Boeing's CFO, Michael Sears, got four months for his role. CEO Phil Condit resigned, and Boeing paid a $615 million fine. The Pentagon canceled the contract in 2006, tarnishing Boeing's reputation and exposing lax oversight.

The deal's rush, treating the KC-767 as a commercial product despite heavy modifications, fueled the mess. Druyun's revolving-door move wasn't unique, but her blatant bias was. Public outrage soared over wasted taxpayer dollars, amplified by media calling it a major Pentagon scandal. The fallout birthed the KC-X program, leading to the KC-46 Pegasus, which Boeing won in 2011 but still faced $7 billion in overruns.

The Fat Leonard scandal was more like a soap opera. Leonard Glenn Francis, a Malaysian contractor nicknamed "Fat Leonard" for his 350-pound frame, ran Glenn Defense Marine Asia (GDMA), a firm servicing Navy ships in Asia. From the 1990s to 2013, Francis bribed over 30 Navy officers with cash, prostitutes, luxury trips, and $2,000 bottles of cognac. In return, they fed him classified ship schedules and steered vessels to his overpriced ports, where he bilked the Navy for $35 million through fraudulent billing.

The scheme was brazen. Francis wined and dined officers at Michelin-star restaurants, hosted sex parties, and even bribed an NCIS agent to dodge probes. The fallout was catastrophic: compromised ship movements threatened national security, and the Navy's 7th Fleet leadership was gutted. Over 440 people, including 60 admirals, faced scrutiny. Thirty-one were charged, 22 pleaded guilty, and careers stalled as investigations dragged on. Francis, arrested in 2013, pled guilty in 2015 but fled house arrest in 2022, only to be recaptured in Venezuela and sentenced to 15 years in 2024.

The major difference between scandals like this and most other countries is that these make the news and are actually rare. In many other countries, corruption in defense spending is the expectation.

❖ ❖ ❖

## International

Imagine a global chessboard where fighter jets, missiles, and tanks are the pieces, and the U.S. sits as the grandmaster. The U.S. arms export business, a $200 billion-a-year juggernaut, shapes alliances, bolsters economies, and stirs controversy. At its core is the Foreign Military Sales (FMS) program, a tightly regulated system that funnels American weapons to allies. Governed by a web of laws and overseen by multiple agencies, it's both a diplomatic tool and a lightning rod for criticism. Andrew Feinstein's *Shadow World* is one of the major critics. Beyond the bureaucracy, U.S. defense companies like Lockheed Martin and Boeing wield influence, quietly steering American foreign policy through their exports.

The U.S. dominates the global arms trade, commanding 40% of exports from 2019 to 2023, according to the Stockholm International Peace Research Institute. Saudi Arabia, India, and Australia snap up F-35 jets, Apache helicopters, and Patriot missile systems, fueling a market that dwarfs competitors like Russia or China. A realpolitik description would be that it isn't about selling gadgets, it's about projecting US power. Arms exports cement alliances, deter adversaries, and keep America's defense industrial base humming. But it's a double-edged sword. Every deal strengthens ties but risks fanning conflicts or returning as a new threat like the MANPADS provided to the Mujahideen, coming back against American forces in the GWOT.

The Foreign Military Sales program is the backbone of this enterprise. Run by the Department of Defense, FMS is like a government-run arms dealership, brokering deals between U.S. contractors and foreign buyers. The Defense Security Cooperation Agency, the Department of State, and then each service's export managers coordinate pricing, delivery, training, and even spare parts. In 2024, FMS approved $80 billion in sales, from $3 billion in Javelin missiles for Ukraine to $10 billion in fighter jets for Poland.

A tangle of laws keeps FMS in check. The Arms Export Control Act (AECA) of 1976 sets the ground rules, requiring exports to serve U.S. national security and foreign policy goals. The International Traffic in Arms Regulations (ITAR) controls sensitive tech, like missile guidance systems, to prevent leaks to adversaries. The Foreign Assistance Act ties exports to human rights, barring sales to countries with egregious violations, though waivers, like those for Saudi Arabia, spark debate. Congress plays referee, with the

Senate Foreign Relations Committee and House Foreign Affairs Committee reviewing deals over $14 million.

Multiple agencies enforce this framework. The State Department's Bureau of Political-Military Affairs sets policy, ensuring exports align with diplomacy. The Pentagon's DSCA handles logistics, while the Defense Technology Security Administration guards against tech leaks. The Commerce Department chips in for less-sensitive items, like radios. Customs Service and the FBI watch for violations, wielding fines or jail time for rogue exporters. This bureaucracy, while clunky, attempts to keep the system honest.

Feinstein's *Shadow World* doesn't pull punches. He argues U.S. arms exports fuel corruption and instability, citing Saudi Arabia's use of American bombs in Yemen's civil war, which killed thousands of civilians by 2024. He slams the revolving door of Pentagon officials joining defense firms prioritizing profits over ethics. Feinstein also points to bloated pricing, like $10,000 toilet seats, as evidence of unchecked greed.

The other side would be that arms exports are just one avenue for conducting foreign policy. Selling F-35s to Japan or HIMARS to Ukraine strengthens allies against China and Russia. Congress also has its eyes on keeping US production lines warm, like Boeing's St. Louis plant. Exports drive economies, fund R&D, and counter international competitor influence.

John Perkins' *Confessions of an Economic Hitman* (2004) is a gripping exposé of how U.S. economic influence subjugates developing nations. Perkins' narrative, blending guilt and insider scoops, paints the foreign economic development tools of the US government as corporate-driven U.S. foreign policy. Economic development, via loans from the former USAID or the World Bank, is one flavor of tying nations to U.S. interests through debt or trade. Arms exports, managed through the Foreign Military Sales program, wield hard power.

When Lockheed pitches F-16s to Taiwan, it's not just a sale; it's a signal to Beijing. Boeing's $20 billion deal with Saudi Arabia in 2023 reinforced U.S. influence in the Gulf, countering Iran. Companies lobby hard, using trade shows like AUSA or Farnborough to woo foreign buyers.

What is unique about the American approach is how much the government relies on private companies to shape foreign policy. There is a quiet shaping of opportunities as defense primes try to align with U.S. goals, like arming Ukraine. Reading the current administration and following the foreign policy objectives is proving to be extremely difficult.

The U.S. arms export business is a paradox. FMS and its laws, from AECA to ITAR, keep it disciplined, with agencies like DSCA and the State steering the ship. As the world's top exporter, America uses the "Fifth Estate" of the defense industry as a tool to accomplish national security objectives. This is an aspect of the defense industry that heightens the importance and makes it stand alone compared to many other consumer industries.

# CHAPTER 9

# Nobody Plays Nice

## Competition, Intellectual Property, and Intelligence Gathering

THE LANDSCAPE OF COMPETITION IN the U.S. defense industry has been shaped by significant technological advancements, geopolitical shifts, and evolving economic realities. This competitive environment has led to some of the most famous and impactful industrial contests in American history, particularly during World War II and the aerospace boom of the 1950s and 1960s. Additionally, as many defense contractors became publicly traded companies, they faced the challenge of competing not only for government contracts but also for investor attention and capital.

World War II marked a turning point in the scale and intensity of competition within the defense industry. The urgent need for advanced weaponry, coupled with the government's willingness to fund ambitious projects, created a fertile ground for innovation and rivalry among defense contractors. One of the most notable competitions during this era was the race to develop the atomic bomb. While this was not a traditional contract competition between vendors, it showcased how the U.S. defense apparatus united for a common goal, contrasted with the company-versus-company rivalries occurring at that time.

Another significant competition during World War II took place in aircraft production. Companies such as Boeing, Douglas, and Lockheed vied for contracts to produce bombers and fighters in unprecedented numbers. The B-17 Flying Fortress, produced by Boeing, and the B-24 Liberator, made by Consolidated Aircraft, emerged from fierce competition and became the backbone of the U.S. strategic bombing campaign. These competitions not only determined which companies would secure lucrative contracts but also influenced the outcome of the war and the future of aviation technology.

The end of World War II did not slow the pace of competition in the defense industry; if anything, the onset of the Cold War and the space race intensified it, particularly in the aerospace sector. The 1950s and 1960s featured some of the most significant contract competitions in the history of the U.S. defense industry.

One of the most prominent contests during this time was the race to develop the first intercontinental ballistic missile (ICBM) for the U.S. Air Force. This competition primarily involved Convair and Martin, with Convair ultimately winning with its Atlas design. The stakes were enormous, not only in terms of contract value but also in shaping the strategic balance of the Cold War.

The 1960s brought even more high-profile competitions. The bid to develop the Tactical Fighter Experimental (TFX) program, which aimed to create a fighter aircraft for both the Air Force and Navy, was particularly contentious. Boeing was initially selected over General Dynamics, but controversy surrounding the selection process led to congressional hearings and a reversal of the decision, ultimately resulting in General Dynamics producing the F-111.

Perhaps the most famous aerospace competition of this era was the race to the moon. While NASA led this effort, it relied heavily on defense contractors. The competition to build the lunar module was particularly intense, with Grumman ultimately emerging victorious over competitors like Boeing and Martin Marietta. This contest not only determined which company would play a crucial role in one of humanity's greatest achievements but also set the stage for decades of space-related contracts.

As the defense industry evolved, many of the major players became publicly traded companies, introducing a new dimension to competition. These firms now had to compete not just for government contracts but also for investor capital and confidence. This shift has had profound implications for how defense companies operate and strategize.

Defense companies competing for investors face unique challenges. Unlike many other industries, their primary customer is the government, and their fortunes are closely tied to political decisions and global events. This reliance can make them appear less predictable and more risky to some investors. However, the stability of government contracts and the essential nature of defense spending can also make these companies attractive, particularly during times of geopolitical tension.

To appeal to investors, defense companies often emphasize their diverse portfolios of products and services, which can help insulate them from shifts in specific defense programs. They also highlight their research and development efforts, showcasing potential for future growth through technological innovation. Many firms have expanded into adjacent markets such as cybersecurity, space technology, and commercial aviation to demonstrate growth opportunities beyond traditional defense contracts.

Investor relations for defense companies also require careful management of public perception. These firms must balance the need to project strength and competitiveness in their core defense business with the growing investor interest in environmental, social, and governance (ESG) factors. Some defense companies respond to this interest by emphasizing their contributions to national security, ethical business practices, and initiatives in areas like energy efficiency and diversity.

Financial performance metrics are crucial in attracting investors. Defense companies often highlight their order backlogs, providing visibility into future revenues. They focus on profit margins and cash flow to demonstrate their ability to operate efficiently and return value to shareholders through dividends and stock buybacks.

The cyclical nature of defense spending poses another challenge in attracting investors. Defense companies often try to demonstrate their ability to weather downturns in military spending by diversifying their customer base internationally or expanding into commercial markets. They may also emphasize their involvement in long-term, multi-year defense programs that can provide stability even during lean-budget years.

The competition for both government contracts and investor capital has shaped the modern U.S. defense industry. From the high-stakes aircraft production contests of World War II to the space race of the 1960s, and now to the complex world of public markets, defense companies have had to continually adapt their strategies. As they look to the future, these firms must balance the demands of their government customers with the expectations of institutional investors while navigating a complex geopolitical landscape and a rapidly evolving technological environment. There is competition between the companies, and then the larger competition as a component of national power on the world stage.

❖ ❖ ❖

## The Shrinking Industrial Base: Consolidation and Its Effects

The defense industrial base network supplying tanks, jets, and bullets, has been shrinking for decades. This squeeze threatens competition, increases costs, and slows innovation. Recent government commissions, like the National Defense Industrial Strategy, are at least describing this problem. It's a high-stakes fix and the clock's ticking.

The DIB's decline is brutal. Since the 1990s, major defense contractors dropped from 51 to five: Lockheed Martin, Boeing, RTX, General Dynamics, and Northrop Grumman. Small suppliers, the supply chain's backbone, are fading too, with a 43% drop in small businesses winning DOD contracts from 2011 to 2020, according to the Government

Accountability Office. The 1993 "Last Supper" consolidation push, post-Cold War budget cuts, and slim margins are not new problems. Complex compliance and a looming cyber-security protocol are making participation in the defense industry harder for smaller or adjacent firms.

Fewer firms mean less price pressure and slower innovation. The DOD's 2022 competition report notes three sources supply 90% of tactical missiles. The top 20 vendors grab half of the DOD's discretionary R&D and procurement opportunities. Less competition also stalls tech leaps when we only have a couple of companies capable of doing an entire platform design and production effort.

DOD and Congressional commissions and committees get assembled every couple of years to examine the problem. The 2024 National Defense Industrial Strategy (NDIS) demands a robust DIB to counter China's fivefold faster weapons buildup. One of the most exciting efforts to address this challenge is the new FoRGED Act championed by Senator Wicker. The Fostering Reform and Government Efficiency in Defense (FoRGED) Act aims to tackle these issues with sweeping reforms. Introduced in December 2024 as S. 5618, the act is paired with Wicker's "Restoring Freedom's Forge" report, targeting inefficiencies in defense acquisition to boost competition and welcome new entrants.

The FoRGED Act proposes a five-pronged fix. First, it slashes red tape by repealing outdated Title 10 regulations and setting five-year sunsets on reporting requirements, freeing up DOD staff. Second, it boosts innovation by making commercial procurement the default, easing entry for tech firms. Third, it fosters competition by simplifying contractor qualifications and mandating prototype competitions with independent down-selects. Fourth, it empowers portfolio acquisition executives over program officers for faster decisions. Fifth, it modernizes budgeting, adopting 2023 PPBE Commission reforms to speed funding.

To broaden the supply chain, FoRGED aligns with the DOD's Office of Strategic Capital, launched in 2023 to offer loans targeting 100 new DIB entrants by 2026. The Defense Innovation Unit is providing another "front door" to fast-track nontraditional players, with $1 billion in 2024 OTA contract. Partnerships with tech hubs, like Austin's Capital Factory, aim to pull in Silicon Valley talent for dual-use tech.

❖ ❖ ❖

## The Competition in Contracting Act of 1984: A Turning Point

The Competition in Contracting Act (CICA) of 1984 is one of the landmark legislation changes in the modern era of the defense industry. Designed to spark competition among suppliers, CICA reshapes how defense companies play the game, and new Trump Administration executive orders in 2025 are cranking up the heat.

CICA emerged from a 1980s push to curb wasteful government spending. Before it, sole-source contracts were common, arguably preventing any real competitive process. CICA flipped the script, mandating "full and open competition" for federal contracts unless exceptions, like urgent national security needs, apply. It required agencies to publish solicitations, use sealed bids or competitive proposals, and justify any non-competitive awards. For the DOD, this means spelling out needs clearly on platforms like SAM.gov, letting all qualified suppliers throw their hats in the ring. Agencies must also debrief losers, explaining why they didn't win, which sharpens future bids.

With CICA there is an entire process of re-competes as programs move through the acquisition lifecycle. Big players face pressure from smaller firms, like Anduril, which nabbed a $250 million drone contract in 2024. The 2022 DOD competition report notes that CICA fosters cost savings with competitive bids, cutting prices by up to 20%. CICA also brought in structure to the bid protest process.

April 9, EO 14266, "Modernizing Defense Acquisitions," demands a DOD overhaul, prioritizing commercial solutions and OTAs over traditional bids. OTAs, which bypass CICA's full competition rules, let firms like SpaceX snag rapid contracts for satellite tech. The April 15 EO, "Restoring Common Sense to Federal Procurement," orders a FAR rewrite to strip non-essential rules, aiming to ease entry for new suppliers. Another April 16 EO pushes commercial products, forcing agencies to justify custom gear within 60 days.

The impact on competition will be a constantly moving dynamic as these new tools are refined in the rulemaking process and industry will take different approaches. New entrants, lured by simpler rules, could lower costs and spur innovation, but primes are thoroughly connected to the Congressional appropriators.

❖ ❖ ❖

## Operation Ill Wind: Exposing Corruption and Reforming the System

One of the largest scandals on competition in the defense industry was an FBI sting called Operation Ill Wind. Launched in 1986 and exploding into public view in 1988, this three-year probe nabbed over 60 contractors, consultants, and government officials for bribery, fraud, and bid-rigging. It shook the Pentagon, birthed tougher laws, and reshaped how defense companies compete.

It all started with a whistleblower. In 1986, a Virginia defense contractor was offered a shady deal: proprietary bid info from a rival for cash. Instead of biting, he tipped off the FBI and Naval Investigative Service (now NCIS). The feds, smelling blood, wired him up, kicking off a probe that snowballed. By June 1988, agents executed 36 search warrants across 12 states, raiding offices of giants like Boeing and Unisys. Wiretaps captured damning chatter about bribes and stealing trade secrets. The investigation ended with a mountain of financial records and guilty pleas, with $622 million in fines and restitution.

Paisley, a former Assistant Secretary of the Navy, was the big fish, pocketing hundreds of thousands in bribes from Unisys and others. He pled guilty, serving four years. Victor Cohen, a deputy Air Force secretary, took London theater tickets and a $17,800 Mercedes payoff, landing a 20-year sentence. Unisys, a major player at the time, paid a record $190 million fine for slush funds and illegal campaign contributions. Over 60 others including consultants, executives from GE and Teledyne, and a Navy procurement officer faced convictions.

Companies paid for bid data to outmaneuver rivals, rigging competitions for contracts like the Navy's submarine-detection systems. Congressional hearings took over the newscycle. Media, like CNN Money in 1993, called it the biggest procurement fraud probe ever.

The fallout reshaped the defense industry's competitive process. Congress, spurred by outrage, passed the 1988 Procurement Integrity Act, a direct response to Ill Wind. It banned procurement officials from sharing bid info or taking contractor jobs for a year post-government, tightening the revolving door. It also mandated transparency in acquisitions, requiring clear justifications for sole-source contracts. The act, amended in 1996, remains a cornerstone document, ensuring "full and open competition" under the 1984 Competition in Contracting Act. These rules forced companies to compete on merit.

Competition got fiercer but fairer. Post-1988, CICA's bid protest process let losers challenge awards, giving some amount of recourse. Trump's 2025 executive orders, like EO 14266, build on this. By pushing OTAs and commercial solutions, the intent is for the DOD to stop generating internal self-development work and truly rely on private industry.

❖ ❖ ❖

## Contract Protests as a Competitive Strategy

In the high-stakes world of platform competition, where billion-dollar deals can make or break a company's future, contract protests have emerged as a powerful and controversial competitive strategy. This process, designed to ensure fairness and transparency in government procurement, has evolved into a complex chess game where companies strategically challenge contract awards to gain competitive advantages.

The mechanics of filing a protest are relatively straightforward, but the implications can be far-reaching. When a company believes that a contract award was improper or unfair, it can file a protest with the Government Accountability Office. Protests can be filed at various stages of the procurement process, from challenging the terms of a solicitation before bids are submitted to contesting the award after a decision has been made.

To file a protest, a company must submit a detailed written argument outlining why they believe the procurement process was flawed. This might include claims of bias in the evaluation process, misinterpretation of proposal requirements, or failure to follow stated evaluation criteria. The GAO then has 100 days to issue a decision, during which time the contested contract is typically put on hold, a period known as the "automatic stay."

While the protest system was created to ensure fairness, many companies have recognized its potential as a competitive tool. The strategic use of protests by companies has become increasingly sophisticated. Some firms use protests as a way to delay a competitor's contract start, buying time to improve their own position or gain supporters in Congress. Others use the process to gain insight into the government's decision-making process, gleaning valuable information about their competitors' proposals or the agency's priorities.

In some cases, companies file protests knowing they are unlikely to win, simply to force the government to reassess its decision or to send a message about their willingness to aggressively defend their market position. This strategy is a feature of this oligopolistic market where a small number of large firms compete for major contracts.

The impact of contract protests on procurement timelines and costs can be substantial. The automatic stay provision means that critical defense programs can be delayed for months while the GAO reviews the protest. In cases where the GAO sustains a protest, the delays can stretch into years if the agency is forced to rebid the contract. These delays not only impact the DOD's ability to field new capabilities but can also result in significant cost increases as programs are extended or reworked.

The threat of protests has led many agencies to be extremely cautious in their procurement processes, leading to overly bureaucratic and time-consuming evaluations as they try to "protest-proof" their decisions. This caution actually drives the DOD to bland, legalistic contract structures that attempt to maximize participation access, but may not truly chase the best technology available.

The most famous recent protest was Palantir Technologies staring down the U.S. Army. The company's bold move, suing its own customer and winning, forced a rethink of how the Pentagon buys tech, opening doors for commercial players and reshaping competition.

The DCGS-A program was the Army's attempt to build a high-tech intelligence platform, pulling data from drones, satellites, and sensors to give soldiers a battlefield edge. By 2015, it was a mess, $3 billion spent over a decade, with soldiers in Afghanistan calling it clunky and useless. Enter Palantir, a data analytics firm co-founded by Peter Thiel, whose Gotham Platform was already helping Special Operations and Marines crunch data fast. Soldiers loved it and it was truly a commercial off-the-shelf platform. The Army's December 2015 solicitation for DCGS-A Increment 2 favored a custom-built system, sidelining commercial options like Palantir's.

Palantir didn't take it lying down. In May 2016, they filed a protest with the Government Accountability Office, arguing the Army's solicitation violated the 1994 Federal Acquisition Streamlining Act, which requires procurement of commercial offerings when available. The Army ignored Gotham's proven track record, pushing a risky, expensive in-house project. GAO shot down the protest, saying the Army's market research was fine. Palantir escalated to the U.S. Court of Federal Claims in June 2016, seeking an injunction to halt the solicitation. Their lawyers, Boies Schiller Flexner, called the Army's approach "illegal and irrational," citing soldier requests for Palantir and a 2015 MITRE analysis suggesting a hybrid commercial solution.

The lawsuit was a serious gamble. Suing a customer, especially the Army, risks burning bridges in an industry where relationships are gold. Palantir's team, backed by Rep. Duncan Hunter's vocal support, bet on a bigger prize: cracking open the Pentagon's walled garden for Silicon Valley. The Army fought back, insisting DCGS-A's unique needs demanded a bespoke system. But court documents revealed cracks with Operational Needs Statements from 2015, which praised Gotham as a "proven capability" meeting all requirements. The Army knew Palantir's potential but plowed ahead, a move the court later deemed "arbitrary and capricious."

In October 2016, Judge Marilyn Kaman ruled for Palantir. She found the Army violated FASA by not seriously exploring commercial options and ordering a new bidding process that included off-the-shelf products. The injunction stopped the Army's $206 million contract, forcing a do-over. Palantir wasn't guaranteed the win, but the ruling put Gotham back in the game. The decision rippled beyond DCGS-A, signaling agencies must justify shunning commercial tech, a nod to broader innovation pushes by then-Defense Secretary Ash Carter.

In March 2018, the Army picked Palantir and Raytheon to develop DCGS-A's next phase, an $800 million deal by 2019. Palantir's Gotham, paired with user-friendly analysis tools, addressed tactical-level requirements. The company's debut at AUSA's 2018 conference, with a slick booth, marked its shift from outsider to player. By 2024, Palantir will have the highest market cap of any defense firm. Dr. Alexander Karp's book *The Technologic Republic* plants their position firmly as Silicon Valley participating and leading national security.

For defense industry executives, understanding the nuances of the protest process and its strategic implications has become a crucial part of competitive strategy. For government procurement officials, navigating the complex landscape of potential protests has become an essential skill in managing major acquisitions. For policymakers, the challenge remains to maintain a system that ensures fairness and accountability in defense procurement while minimizing unnecessary delays and costs.

The use of contract protests as a competitive strategy in the US defense industry underscores the complex interplay between government, industry, and oversight bodies in the defense acquisition process. It reflects the high stakes involved in defense contracting and the intense competition for these lucrative and strategically important contracts. As the industry continues to face new challenges and opportunities, the protest process will undoubtedly continue to play a crucial role in shaping the competitive landscape of US defense procurement.

❖ ❖ ❖

## Corporate R&D and IP Crown Jewels

Independent Research and Development (IRAD) in the defense industry is a unique accounting feature where companies fund their own offerings as opposed to government-funded R&D, where the project is contractually aligned. IRAD is where the companies try to "skate where the puck is headed" and have their own offerings. The Lockheed Martin Skunk Works facility and Boeing's Phantom Works are two of the most famous hubs. Unlike commercial R&D, IRAD is reimbursed through overhead costs on government contracts, blending private innovation with public funding. The battle over IP data rights is one of the hardest aspects of negotiating new contracts.

IRAD is a peculiar creature. Unlike pure private R&D, where firms like Apple fund their own iPhone breakthroughs, IRAD lets defense contractors recover costs via DOD contracts, as long as projects align with military interests. According to the National Defense Industrial Association, the top five primes spent $12 billion on IRAD in 2023. Chasing tech like hypersonics or quantum sensors. It's not a blank check but allows companies to fund the research they want to conduct and then wrap that cost into their overhead fees spread across their contracts. This structure arguably favors existing primes that have many contracts to spread this cost across.

Contractors lean on DOD signals like budget priorities or the latest threat briefs. Lockheed's Skunk Works, famed for the SR-71, targets high-risk, high-reward projects, betting on game-changers like sixth-gen fighters. Software firms like Anduril and Palantir focus on dual-use tech, like AI software, to hedge bets for commercial markets. Profit margins, capped at 10-12%, push firms to chase long-term sustainment contracts, where real money flows. It's a chess game, invest where the Pentagon's headed, but don't bet the farm.

The IP data rights battle is where things get heated. *Formation of Government Contracts* explains the Defense Federal Acquisition Regulation Supplement framework: contractors can claim "limited rights" for privately funded tech, restricting government use, or face "unlimited rights" if the DOD funds development, granting broad access. The Pentagon, wary of vendor lock-in, pushes for unlimited rights to tweak or repair systems without relying on primes. A 2024 War on the Rocks piece noted this stifles innovation, firms like Palantir hesitate to share AI code if the government can hand it to rivals. Industry, meanwhile, guards IP to protect commercial spin-offs or future contracts, as seen in Boeing's 2020 court win to mark unlimited rights data as proprietary to hide their technology descriptions from FOIA disclosures to competitors.

The new administration has been on both sides of the issue. Executive orders have encouraged a major pivot to truly embracing commercial technology, but then also demanding government rights like the "right to repair." There is a constant battle between the DOD asking companies to show up to demonstrations and events with working hardware (funded at largely private expense) and then wanting IP rights and technical data that companies fear could just be handed off to competitors and told to build a copy.

Contractors chase DOD priorities, balancing risk and reward, while the government pushes for data control to avoid lock-in. This dynamic of IRAD is another example of the DOD pushing risk onto private industry and defense contractors absorbing risk in order to remain in position for the next opportunities.

❖ ❖ ❖

## Competitive Intelligence in the Defense Industry

The function of competitive intelligence (CI) in the defense industry is similar to other industries, but has a unique aspect of access to customer information through Freedom of Information Act requests. It involves the process of gathering, analyzing, and acting on info to outsmart rivals. CI shapes how companies bid, innovate, and survive in a cutthroat market. It's governed by laws like the Economic Espionage Act, relies on tools for open-source intelligence (OSINT), and leans on processes like the Black Hat review to sharpen strategies. Competitive intelligence is the defense industry's crystal ball. It's about understanding rivals' strengths, weaknesses, and moves while predicting Pentagon needs. CI informs bids, helping firms craft proposals that undercut competitors on price or dazzle with tech.

The Economic Espionage Act of 1996 is the main structure. It criminalizes stealing trade secrets, with penalties up to seven years in prison or $5 million in fines. Defense firms, guarding proprietary data like missile specs, face risks from competitors but the main risks are from foreign intelligence collectors. The act, paired with the 1988 Procurement Integrity Act, bans sharing bid info, a lesson from the Operation Ill Wind scandal.

For open-source intelligence, defense companies are like detectives scouring public clues. They tap platforms like SAM.gov for RFP details or Jane's for rival tech specs. Social media, especially LinkedIn, reveals competitor hires. Trade shows like AUSA yield gold: a rival's booth demo or exec's offhand remark can signal strategy.

The Black Hat process and determining "Price to Win" is where a defense contractor's CI team really earns their pay. The Black Hat brainstorms how rivals will pitch, price, or partner, then stress-tests the home team's proposal.

❖ ❖ ❖

## International Competition and Collaboration

Competition between U.S. defense industry companies and international firms is a minor aspect compared to how the US defense industry faces foreign intelligence agencies. The foreign agents aren't chasing Pentagon gossip or diplomatic cables, they're after the drawings and material component lists of US weapon platforms. For foreign governments, stealing this intellectual property is a key part of their military strategy; it's a shortcut to closing the tech gap without decades of R&D. China's relentless IP theft, from the F-35 to missile tech, has cost billions and reshaped global power dynamics.

Foreign companies, like China's AVIC or Russia's Rostec, and state-backed intelligence collectors are actively funded to seek US platform information. Unlike domestic rivals like Lockheed Martin or Boeing, who compete for contracts, foreign players don't play by the same rules. They're not bidding on SAM.gov, they're hacking servers, dispatching spies, and stealing equipment from third-party nations. A 2024 FBI report estimates 80% of economic espionage cases target defense firms, with China leading the pack. In *The Kill Chain* (2020), Christian Brose (former SASC staff and now President of Anduril) outlines China's aggressive strategy to steal American military secrets, aiming to erode U.S. technological dominance. The Chinese Communist Party prioritizes cyber espionage, targeting defense contractors like Lockheed Martin to steal weapon designs. Brose details how state-backed hackers, like the APT10 group, infiltrate networks to swipe terabytes of data, avoiding billions of dollars of their own investment and catching up. These thefts fuel China's J-20 and J-31 jets, mimicking U.S. capabilities. Beyond cyber, China exploits open-source intelligence, scouring trade shows like AUSA and platforms like LinkedIn for tech insights.

Brose argues China's focus is on weapon specs like AI, hypersonics, and drones over old-fashioned Department of State communication messages. The CCP's "Made in China 2025" plan drives this, blending civilian and military tech theft, like GE Aviation's C919 data. The Economic Espionage Act of 1996, with its seven-year penalties, struggles to deter state-sponsored hacks. Brose notes China's edge: a unified strategy exploiting U.S. bureaucratic silos and lax contractor cybersecurity.

Foreign governments prioritize weapon designs over U.S. government secrets because the former are harder to replicate. Internal secrets like the location of ships and military units change fast and lose value. But a fighter jet's radar-absorbing material or a missile's guidance code can catch up their capability in only months.

China's IP thefts are the stuff of nightmares for U.S. defense firms. The 2014 F-35 breach was a massive success for

China's state-backed APT10 group. Hackers swiped tera-bytes of data from Lockheed Martin and subcontractors, including stealth tech and sensor designs, costing an esti-mated $50 billion in lost advantage. The breach, exposed in 2015, helped China's J-20 and J-31 jets mimic F-35 features, narrowing the stealth gap. From 2010 to 2015, China's Ministry of State Security targeted GE Aviation and Honeywell, stealing C919 airliner tech, leading to a 2019 DOJ indictment of two Chinese operatives. The 2005 "Titan Rain" attacks, linked to the People's Liberation Army, hit NASA and Lockheed, nabbing Space Shuttle and missile data. These heists arm China's military-industrial complex.

It is not only China and Russia that literally attack US defense companies. Even allies like Israel and France have been found to steal military technology. The book *Friendly Spies* by Peter Schweitzer has some pretty incredible quotes and details about France's history of economic and mili-tary theft. A Wall Street Journal article from 1992 quotes a former Justice Department official who "regarded Israel as being the second most active foreign intelligence service in the United States." In 1996 the GAO issued a report stating that Israel "conducts the most aggressive espionage oper-ations against the United States of any ally.... routinely resorts to state-sponsored espionage [to steal] classified military information and sensitive military technology."

This battle isn't abstract. It's about data security for servers in the back room of a machine shop. Defense contractors are on the front lines against threats and allies, trying to keep any military advantage we may have. Protecting IP isn't a corporate game; it's a national imperative where every stolen file matters.

# CHAPTER 10

# Congress Gets a Vote

*"Remember, write to your Congressman. Even if he can't read, write to him."* —Will Rogers

WHEN WE THINK ABOUT THE U.S. defense industry, it's tempting to picture sleek fighter jets or sprawling shipyards humming with activity. But the real engine behind this extra-national power is Congress. It's where budgets are born, acquisitions are scrutinized (mostly for theatric effect), and the balance between national security and corporate influence plays out. The House and Senate are like the Great and Powerful Oz behind the curtain. They shape, incentivize, publicly shame, and force an industry where the U.S. government stands as the sole customer, The stakes are high: every decision ripples through global security, local economies, and the lives of countless workers. From the F-35 program's ballooning costs to the Navy's shipbuilding debates, Congress doesn't just oversee the defense industry; it defines its very contours.

The story begins with the acquisition, where lawmakers craft the rules governing how the Pentagon buys everything from drones to destroyers. Committees like the House Armed Services Committee call witnesses from major primes, balancing cutting-edge tech against taxpayer wallets. It's a messy process, think of herding cats with billion-dollar collars. Appropriations, meanwhile, are where the money flows. The annual National Defense Authorization Act is Congress's love letter to the Pentagon dictating funds for jets, ships, and missiles. But authorizers just decide to throw a party, and the appropriators decide whether it will be Boone's Farm or Pappy Van Winkle. Surprisingly no one, it is not all numbers and spreadsheets. Local and regional politics sneak in. Senators fight for contracts that keep jobs in their states, turning defense spending into a hometown issue. That's Congress at work, blending strategy with sentiment.

Then there's lobbying, the shadow player. Defense giants spend millions to sway Capitol Hill, their influence woven into campaign contributions and closed-door meetings. It's a world of high stakes and bad banquet food. Congress is allegedly the watchdog, holding hearings to probe cost overruns or contractor missteps. As the new administration shapes new government policy with posts on social media, Congress definitely has a challenge in determining what is appropriate advice and consent. Two of the best books to describe the Congressional / Pentagon dynamic are *The Pentagon's Wars* by Mark Perry (yes, different from *The Pentagon Wars)*, a gripping look at congressional-Pentagon dynamics, or *The Power Game* by Hedrick Smith, which dissects Washington's influence machine. Both capture the messy, human heart of defense policymaking.

❖ ❖ ❖

## Authority and Dunning-Kruger

The U.S. Congress exerts profound influence over defense acquisition, shaping the processes through which the DOD procures technology and services. The U.S. Constitution grants Congress several authorities that directly shape the Department of Defense and the defense industry. Article I, Section 8 empowers Congress to "raise and support Armies," "provide and maintain a Navy," and "make Rules for the Government and Regulation of the land and naval Forces." These clauses establish Congress's role in creating, funding, and overseeing the military, including the DOD's structure and operations. The power to "declare War" and regulate "Captures on Land and Water" is a uniquely American experiment attempting to place this control in the most democratically elected segment of the government. Congress's authority to "provide for organizing, arming, and disciplining " is a unique check on otherwise significant power assignments to the executive branch. The appropriations power, rooted in the clause requiring all funds to be drawn from the Treasury, gives Congress control over the DOD's budget through the National Defense Authorization Act and appropriations bills. Additionally, the commerce clause enables regulation of defense contractors, while the necessary and proper clause supports laws like the Federal Acquisition Regulation. These powers collectively enable Congress to shape defense policy, acquisitions, and industry dynamics, ensuring alignment with national security and fiscal priorities

Congress exerts its significant authority over the defense industry in very unpredictable ways. This influence shapes the procurement of weapons systems, allocation of funds, and economic impacts of defense spending. Congressional committees, budget processes, and interventions in acquisition programs show occasional national security, fiscal responsibility, and regional interests. But often struggles to avoid capricious incompetence.

Congressional committees, notably the House Armed Services Committee (HASC) and Senate Armed Services Committee (SASC), play a pivotal role in crafting acquisition policies. These committees draft the NDAA, which authorizes defense programs and includes provisions to refine procurement processes, such as promoting competition or accelerating technology adoption. Through oversight hearings, HASC and SASC scrutinize DOD proposals, ensuring alignment with strategic priorities and cost estimates.

Before the DOGE era, Congress had already instituted its killswitch Nunn-McCurdy Act. Enacted in 1982, it is a critical tool in the Congressional oversight toolbox. The law was designed to curb cost overruns in major defense acquisition programs. Named after Senators Sam Nunn and David McCurdy, it requires the DOD to monitor and report significant cost growth in major defense acquisition programs. The law establishes two thresholds: a "significant" breach (cost growth of 15% or more over the baseline) and a "critical" breach (25% or more). Upon a breach, the DOD must notify Congress, and for critical breaches, the program faces termination unless the Secretary of Defense certifies its necessity, cost control measures, and management restructuring within 60 days. The F-35 program triggered a

Nunn-McCurdy breach in 2010, prompting restructuring to address costs exceeding $400 billion.

Balancing innovation, cost, and oversight remains the same challenge that General Washington and his quartermasters faced in the 1780s. The DOD seeks advanced technologies, such as autonomous systems and cyber defenses to counter adversaries. Innovation and invention carry high, unknown costs. The Government Accountability Office and Congressional Research Service help inform members on the details and accounting of DOD programs.

The allocation of defense funds follows a structured process (although Congress has barely been able to actually do it in over 20 years). The DOD submits a detailed budget request outlining needs for weapons, research, and operations. Congressional committees review this request, informed by Pentagon briefings, GAO analyses, and industry input. Congress sets procurement guidelines, such as "fly before you buy" testing, to supposedly mitigate risks.

Congressional priorities, particularly regional economic interests, significantly influence budget decisions. Defense contracts sustain jobs, with the F-35 program being one of the most famous successes in spreading the peanut butter. The Joint Strike Fighter supports employment across 45 states, and this was not done accidentally and certainly not for efficiency. Lawmakers advocate for projects benefiting their constituencies, such as shipbuilding in Virginia or missile production in Alabama, balancing economic and security objectives. While this ensures broad political support for defense budgets, it can lead to funding for programs of questionable strategic value.

Budgetary disruptions, such as continuing resolutions (CRs) and sequestration, pose significant risks to the defense industry. CRs are enacted when Congress fails to pass timely budgets, freeze spending at prior levels, halting new contracts and delaying programs. The previous twenty years of mostly CRs have been topped by the full year Fiscal Year 2025 CR that somehow still allows significant reprogramming. Sequestration, triggered by the 2011 Budget Control Act, imposed $500 billion in defense cuts over a decade. Smaller contractors, reliant on steady cash flow, were particularly hard-hit.

In an aspirational dream Congress would be up to the challenge of evaluating US military strategy and balancing it with economic realities. Unfortunately, it is more of a Dunning-Kruger effect in action with billions of dollars involved. The Dunning-Kruger effect is a cognitive bias where people with low skill or knowledge in a specific area overestimate their competence, while those with high skill may underestimate theirs. It happens because beginners lack the awareness to recognize their gaps, leading to inflated confidence, whereas experts, aware of complexities, feel less certain. For example, an amateur might boldly claim expertise in defense policy, unaware of nuances, while a seasoned analyst might doubt their grasp of every detail. The new Administration is embracing inexperienced appointees as a means to "drain the swamp." Hopefully this approach will turn into valuable changes that actually improve Congress' ability to "provide for the common defense."

❖ ❖ ❖

## The Other Rich Men North of Richmond

K Street in Washington, D.C., is more than a stretch of pavement; it's the nerve center of American lobbying, where power suits and sharp pitches collide to shape policy. Lined with sleek offices, this corridor hums with firms representing every industry, from tech to pharmaceuticals and of course the Fifth Estate of national power. Here, lobbyists who are often ex-lawmakers or former Pentagon officials, craft strategies to sway Congress. In 2023, lobbying spending topped $4 billion. K Street's influence isn't just about money; it's about access, relationships, and knowing which doors to knock on. For the defense industry, where the U.S. government is the sole customer, K Street is the battlefield for securing contracts and shaping budgets.

This consequential game targets Congress with precision. Defense primes spend millions ($150 million in 2023) to influence the House and Senate Armed Services Committee Members. Lobbyists, often former Hill staffers, slip into offices with briefs on job creation or threats from China, building cases over years of coffee chats and gala dinners.

Money amplifies their voice, and Political Action Committees (PACs) are the megaphone. In 2022, Lockheed Martin's PAC dropped $2.8 million, mostly on defense committee members. Boeing and RTX followed, spreading donations across parties to ensure access. These aren't payoffs; they're keys to the room. A senator backed by Northrop Grumman's PAC may be more likely to hear their pitch on B-21 bombers.

Beyond cash, firms wield the jobs card, F-35's 180,000 jobs across 45 states or the extremely regional priorities of shipyards on the coasts. This economic leverage turns budget talks into local stakes as the Space Command relocation football keeps kicking between Colorado and Alabama.

Ethics rules try to keep things clean. The Lobbying Disclosure Act mandates the reporting of expenses and clients, while the Foreign Agents Registration Act flags foreign ties. While these may be aspirational objectives, even accused and allegedly compromised Representatives like Eric Swalwell are able to remain in office and continue to be re-elected.

The defense lobby practice on K Street has serious representation. Firms like Brownstein Hyatt Farber Schreck, JA Green, and Cassidy & Associates are titans in this arena. Each brings their unique playbooks of bipartisan clout, niche expertise, or storied connections to shape policy.

Brownstein Hyatt Farber Schreck is a dominant player in defense lobbying. Founded in 1968 by Norm Brownstein, a Denver lawyer dubbed "America's 101st Senator" for his bipartisan sway, the firm's D.C. office is a juggernaut with over 70 lobbyists, including former Senators and Pentagon insiders. Brownstein's defense roster includes giants like Boeing and General Motors.

JA Green & Company takes a leaner, laser-focused approach. Founded by Jeffrey Green, a former House Armed Services Committee staffer, this boutique firm thrives on deep Pentagon connections and defense expertise. Green's team includes ex-military officers and Hill veterans who know the NDAA's fine print. They represent clients like BAE Systems and smaller tech firms, focusing on niche areas like cybersecurity and unmanned systems.

Cassidy & Associates, once K Street's undisputed leader faces plenty of upstart competition. Founded in 1975, the firm pioneered modern lobbying in the defense industry. With $5.5 million in Q1 2021 revenue, Cassidy's defense clients include Northrop Grumman and shipbuilder Huntington Ingalls.

For the real analysis of K Street lobbying, one of the main texts is *The Lobbyists: How Influence Peddlers Work Their Way in Washington* by Jeffrey H. Birnbaum. This 1993 classic pulls back the curtain on Washington's lobbying machine. Birnbaum, a veteran Wall Street Journal reporter, shadows lobbyists to reveal their tactics of schmoozing lawmakers, crafting narratives, and leveraging connections.

❖ ❖ ❖

## Citizens United Says Companies Get a Vote

Defense companies do not just sit around waiting for a lobbying firm to call about a fundraising dinner. They host some of the largest corporate Political Action Committees in the United States. These groups, fueled by employee and executive donations, pour millions into candidate election campaigns, shaping who holds power and what policies prevail. In the defense industry, where giants like Lockheed Martin and Boeing rely on Congress for contracts, PACs are just a cost of doing business.

PACs weren't always the powerhouse they are today. Born from the 1943 Smith-Connally Act, which banned unions from direct campaign donations, PACs emerged as a work-around for labor to pool funds. Corporations caught on after the 1971 Federal Election Campaign Act formalized their role, letting companies form PACs to collect voluntary contributions from employees. The real game-changer came post-Watergate when FECA's 1974 amendments capped individual donations but left PACs room to thrive, setting limits like $5,000 per candidate per election. Then, the 1976 Buckley v. Valeo decision struck down spending caps as free speech, opening the floodgates. By the 1980s, corporate PACs were a fixture, and defense firms had only one customer and campaign contributions were the only ticket to ride.

The 2010 Citizens United v. FEC decision supercharged this growth. By treating corporations as "people" with free speech rights, it unleashed unlimited independent expenditures, paving the way for Super PACs. While corporate PACs still face FECA limits, the ruling's ripple effect freed up cash flow, letting PACs coordinate with Super PACs for bigger impact. In 2022, the defense sector's PACs donated $14.8 million, with Lockheed Martin's PAC donating $2.8 million.

Why do defense firms lean so hard into PACs? The Pentagon is their only customer, and Congress controls the purse. A well-placed donation can mean a Senator's ear (or hearings) promoting a major industry capability or at least grandstanding against a potential rival. Defense PACs wield the jobs card, a potent weapon in congressional districts. Lockheed Martin's F-35 supports 180,000 jobs across 45 states, a stat that Lockheed does not hide. In 2022, Northrop

Grumman's PAC donated $1.7 million, with heavy focus on lawmakers backing the B-21 Raider bomber. This isn't corruption in the legal sense, every dollar was reported, but it shows how PACs tilt the scales, ensuring defense priorities align with corporate bottom lines.

◈ ◈ ◈

## All Politics Are Local

Factory parking lots in Georgia are packed with pickup trucks as assembly lines buzz with workers building C-130J cargo planes. This isn't just a workplace; it's a lifeline for a town where every paycheck fuels the local grocery store and keeps the lights on at city hall. Congressional representation makes this possible, turning defense contracts into economic anchors for communities across the U.S. Lawmakers, especially those on the House and Senate Armed Services Committees, don't just debate national security; they champion projects that keep their districts humming.

This hustle often slides into "pork barrel" politics, where defense spending doubles as a political scorecard. Pork barrel is the art of funneling federal funds to pet projects, not always for strategic genius but to win votes back home. The NDAA hosts one of the most maneuverable elements of "discretionary spending" that Congress has access to. Lawmakers slip in funds for bases or weapons the Pentagon doesn't even request. In 2018, Congress approved $1.7 billion for extra C-130Js the Air Force didn't request.

In the small city of Lima, Ohio, where the Joint Systems Manufacturing Center churns out M1 Abrams tanks, the power of congressional representation has kept production lines running even when the Army has begged them to stop. The JSMC is a GOCO facility run by General Dynamics Land Systems. It is the only plant in the U.S. building and upgrading these 70-ton behemoths, and it's a lifeline for Lima's economy. Ohio's lawmakers from both parties have repeatedly overridden Army attempts to halt production. Army acquisition officials joke about Congress wanting "a tank for every driveway."

In 2012, then-Army Chief of Staff Gen. Raymond Odierno testified that the Army, with over 6,000 Abrams tanks, had enough to meet its needs. Facing budget constraints and a shift to lighter forces for counterinsurgency wars in Iraq and Afghanistan, the Army proposed idling the Lima plant from 2014 to 2017 for $3 billion in savings. Odierno argued tanks were less critical in urban combat, and foreign sales could keep the plant afloat. But Ohio members Senators Sherrod Brown (D), Rob Portman (R), and Rep. Jim Jordan (R) found a rare bipartisan moment of agreement.

Brown and Portman, wielding seats on the Senate Armed Services Committee, pushed back hard. Portman, in 2012, secured $255 million for upgrades through 2014, framing it as vital for both jobs and national security. Rep. Mike Turner (R), chair of the House Armed Services Tactical Air and Land Forces Subcommittee, ensured Congress allocated $120 million in 2017 for upgrades the Army didn't request. These funds kept the plant at a steady hum, producing 11 to 15 tanks monthly by 2025. The lawmakers' argument is they are preserving the industrial base to ensure long-term military readiness. The truth is these representatives are not wrong.

This persistence paid off. In 2016, Congress approved $558 million for more tanks, ignoring Odierno's pleas. By 2019, a $714 million contract boosted production, doubling the workforce to 1,000. In 2024, Congress earmarked $2 billion for plant modernization, enabling new hull production for the first time since the 1990s. The "tank for every driveway" joke echoes the concept at the start of this book. Do we have a nation (and its Army) to defend an economy, or do we have an economy to defend our nation. The Lima saga shows pork barrel politics at its peak. Jobs, not just security, drive decisions.

Balancing national security with constituent needs is like juggling flaming torches while riding a unicycle. But lawmakers know voters prioritize jobs now, not abstract threats. During NDAA hearings, a congressman might back a new submarine contract for Electric Boat in Connecticut over a cyber program in Silicon Valley because shipyards employ 10,000, while tech firms hire a fraction. It's not all self-interest; defense spending props up 3.5 million jobs, anchoring communities.

Congress, the economy, and national power collide to shape a vital industry. Lawmakers, wielding powers to raise armies and control budgets, steer the course of defense acquisitions and funding. The sobering truth is that Congress often stumbles, hampered by limited expertise and relying on lobbyists and the defense industry to inform them of many basic facts of realpolitik and international affairs. Defense contractors fill this gap by guiding lawmakers through complex terrain.

The result is a Congress that, while powerful, struggles to balance strategic needs with parochial demands. Rep. Hank Johnson's 2010 suggestion that Guam might "tip over" from additional troops deployed to the island demonstrates how uneven and often incompetent Congress can be in determining things as important as military power.

This complexity underscores why defense contractors lean on lobbying. Many lawmakers, juggling myriad issues, lack the depth to parse hypersonics or cyber defenses. The Dunning-Kruger effect remains a feature of every new Congress inaugurated. K Street actually does provide a service to the country in exchange for their fees.

This chapter leaves us at a crossroads. Congress's power to shape the defense-industrial complex is unmatched, but its reliance on lobbying and regional biases reveals a system creaking under its own weight. In his farewell address on January 17, 1961, President Dwight D. Eisenhower famously warned of the dangers of the "military-industrial complex," a term that has since defined the intricate relationship between the U.S. military, defense industry, and government. However, early drafts of the speech included a more expansive phrase: "military-industrial-congressional complex." This original wording, which explicitly acknowledged Congress's role, was ultimately revised to omit the keyword. This was driven by ongoing legislative agendas that even a sunsetting Presidency couldn't risk by poking Capitol Hill.

# CHAPTER 11

# The Second Oldest Profession – Acquisition Reform

## The Infinite Rules, Regulations, and the Constant Desire to Change Them

*"Acquisition reform is the second oldest profession."*
—LTG (R) David Bassett

SINCE GENERAL GEORGE WASHINGTON, THERE have been continuous calls for everything from budget reviews to structural reforms and occasional hanging. Arguably, no system as large as fielding and maintaining an Army, Air Force, Navy, and now a Space Force will ever be fully efficient. There are too many decisions that force extreme trade-offs in policy and budget space. A topic worthy of a different book would be whether competing stakeholders driving towards compromise (and therefore not committed to a single priority) is a bug or a feature of our defense acquisition process.

*Pentagon Wars* by COL (R) Jim Burton is a tremendously detailed look at the extreme machinations and dedication of some loyal insurgents who took the Bradley Fighting

Vehicle from a failing program that was a rolling death trap and made it into the most capable infantry fighting vehicle for over 35 years. Another series of examples of the internal drive forcing the calls for acquisition reform is well highlighted by Senator William Proxmire's "Golden Fleece Awards" from 1975 to 1988. During his tenure, he "awarded" 168 dubious and likely wasteful expenditures, especially in the research and development fields. His focus was not only on the Department of Defense, but he highlighted some blatant waste from NASA, USPS, and many other departments. However, the US Army spending thousands of dollars researching how to buy Worcestershire sauce sets the tone for the waste that only a faceless bureaucracy can defend by creating checklists to put the checkboxes in order.

❖ ❖ ❖

## Turn of the Century Through World War II

The War Department office in 1899 was still using gas lamps on desks while grappling with the chaos of supplying an army stretched thin by the Spanish-American War. This was the world Elihu Root, newly appointed Secretary of War, stepped into. His reforms would echo through World War II. Root's efforts tackled a recurring problem: how to equip a military efficiently without wasting money or lives. From his foundational reforms to Bill Knudsen's industrial wizardry during WWII, this broad era laid the groundwork for modern defense procurement. Themes of centralization, industry collaboration, and cost control weave through a story of legislation, crises, and colorful characters, shaping an acquisition system that still influences the Pentagon in 2025.

Root's reforms were a response to a mess. The Spanish-American War exposed the War Department's logistical nightmares such as spoiled rations, shoddy rifles, and troops dying of disease before reaching Cuba. Root, a sharp lawyer with no military background, saw the problem wasn't just supply but structure. The Army's procurement was a patchwork of feuding bureaus, each buying its own gear with little oversight. In 1899, Root pushed for centralized control, streamlining purchasing under a single authority to curb waste. His 1901 Annual Report called for a General Staff to coordinate planning, a radical idea that became law in 1903 via the General Staff Act. This act unified acquisition decisions, cutting redundancies like duplicate cannon orders. Root's vision wasn't flashy, but it was a lifeline, setting a precedent for top-down efficiency that resonates in today's Federal Acquisition Regulation.

Fast forward to World War I, and the limits of Root's system showed. The U.S. entered the war in 1917 with an Army smaller than Belgium's, scrambling to arm millions. Procurement was a circus with factories churning out mismatched rifles and ships lacked guns because bureaus and agencies were still extremely parochial. The 1918 Overman Act gave President Wilson near-dictatorial powers to reorganize agencies, centralizing acquisition under the War Industries Board. Led by Bernard Baruch, the board prioritized production, redirecting steel from cars to tanks and syncing industry with military needs. It worked and by the time of the Armistice, the U.S. was actually running airplane production lines, but peacetime dismantled these gains. Congress was wary of centralized power, let the War Industries Board lapse, and the 1920s saw procurement revert to fragmented bureaus.

The interwar period wasn't all stagnation. The 1920 National Defense Act tried to keep the lessons of WWI alive, mandating joint Army-Navy planning and creating the Assistant Secretary of War for procurement. But budgets shrank, and isolationism gripped Congress, leaving the military with outdated gear. By 1939, as Hitler's tanks rolled, the U.S. Army had just 1,800 mostly obsolete planes. The Great Depression forced frugality, but it also birthed ideas like cost-plus contracts, where firms like Curtiss-Wright were guaranteed profits to offset risks of low-volume production. These contracts, formalized in the 1930s, became a double-edged sword, speeding output but inviting cost overruns.

Then came World War II, and the stage was set for Bill Knudsen. He is described as a true maverick reformer in Arthur Herman's seminal *Freedom's Forge*. Knudsen, a Danish immigrant and former GM president, was tapped by FDR in 1940 to lead the National Defense Advisory Commission. Herman paints him as a production maestro, turning America's industrial might into an "arsenal of democracy." Knudsen's reforms were pragmatic, not bureaucratic. Facing a military needing 50,000 planes and 80,000 tanks, he sidestepped red tape. Coming from a Detroit background he was able to convince carmakers like Ford to build bombers and Chrysler to churn out tanks. His "conversion" strategy retooled auto plants for war and by 1942, Ford's Willow Run plant spit out a B-24 bomber every 63 minutes. Knudsen's genius was collaboration, not control. He offered fixed-price contracts to cap costs but sweetened deals with tax breaks, ensuring firms like Boeing stayed profitable while meeting deadlines.

Knudsen's reforms weren't flawless. Early chaos saw shortages of steel and rubber, and labor strikes slowed production. The 1940 Walsh-Healey Act, meant to ensure fair wages, made it extremely difficult for small suppliers. The 1941 Lend-Lease Act was a game-changer, funneling $50 billion in aid to allies and boosting U.S. factories. By 1944, the U.S. produced 40% of the world's weapons, a feat Root could only dream of. Knudsen's approach of partnering with industry, not dictating, set a template for modern public-private partnerships, echoed in 2025's push for commercial tech and silicon chip re-shoring.

The 1940 Second Revenue Act eased taxes on defense profits, spurring investment, while the 1942 War Powers Act gave FDR authority to prioritize contracts, streamlining acquisition. These laws, though temporary, tackled recurring themes: speed versus cost, and government versus industry control. The War Production Board, succeeding Knudsen's office, refined his model using "priorities" to allocate materials. But wartime haste bred waste as cost-plus contracts led to $700 million in overruns on ships alone. These lessons shaped post-war reforms like the 1947 Armed Services Procurement Act, which standardized buying across services.

What ties Root to Knudsen is the struggle to balance efficiency with accountability. Root's centralized General Staff fought bureaucratic sprawl, much like Knudsen's factory conversions battled production bottlenecks. Both faced a Congress skeptical of overreach as Root's reforms needed years to pass, and Knudsen dodged lawmakers' microman-aging. Their era also highlights the industry's growing clout. By WWII, firms like Lockheed were no longer just suppliers but partners, approaching or reaching national prominence and dependence.

This period wasn't perfect. Root's reforms didn't prevent WWI's chaos, and Knudsen's miracles couldn't eliminate waste. From 1899 to 1945, acquisition reform evolved from Root's structural fixes to Knudsen's industrial genius. Key events like the Spanish-American War's failures, WWI's scramble, and WWII's production triumph forced change.

❖ ❖ ❖

# The 1960s

Pentagon offices in the 1960s were filled with typewriters and a permanent fog of cigarette smoke. The Cold War was in full swing, and the U.S. defense budget was ballooning to counter Soviet missiles and Vietnam's jungles. Beneath the urgency, defense acquisition was arguably an abso-lute disaster riddled with cost overruns, shoddy deals, and bureaucratic bloat. Reformers, armed with new laws like the Truth in Negotiations Act worked to bring systems approaches to the constant challenge of acquisition. The whistleblower A. Ernest Fitzgerald exposed some of the absolute worst aspects of a weak defense industry in his scathing book *The High Priests of Waste.* The 1960s were a crucible for acquisition reform, marked by legislative fixes, fierce debates, and systemic flaws still being debated and challenged in new Executive Orders today.

The decade kicked off with a wake-up call. The 1950s had seen defense spending soar, but programs like the B-70 bomber and the Navy's Talos missile bled cash with little to show. By 1960, the Pentagon's budget hit $45 billion but reports of $2,000 toilet seats and mismanaged contracts sparked outrage. The problem wasn't just money; it was process. Procurement was a tangle of service rivalries, still duplicating efforts. Cost-plus contracts, guaranteeing firms like Lockheed profits regardless of efficiency, invited waste.

Robert McNamara was appointed by Kennedy as Secretary of Defense and brought his Ford Motor Company mindset to the Pentagon. McNamara's big idea was systems analysis using data to streamline buying and cut costs. In 1961, he centralized acquisition under the Defense Supply Agency, aiming to unify purchasing across services. This is now being tried again in 2025, by using GSA as a central procurement agency. His Planning-Programming-Budgeting System (PPBS) tied contracts to long-term goals, forcing services to justify every dollar. It sounded smart, but Fitzgerald's *The High Priests of Waste* argued PPBS was a "gimmick," burying real issues under spreadsheets while contractors gamed the system.

Congress, smelling blood, stepped in with the Truth in Negotiations Act (TINA) of 1962, TINA was a direct jab at contractor deceit. It required firms like Boeing and General Dynamics to submit certified cost data during negotiations, ensuring bids weren't relying on "rubber baselines." If costs later spiked, contractors had to justify every penny or face penalties. TINA aimed to level the playing field, especially for sole-source contracts where there was not enough competition.

Fitzgerald's insider view called Pentagon brass and contractors "high priests," worshipping waste in a cult of inefficiency. The F-111 fighter, a McNamara pet project, was a poster child. Designed for both the Air Force and Navy, it ballooned to $7 billion by 1967. Fitzgerald, testifying to Congress in 1968, revealed a $2 billion overrun. He put the concept of contractors "buying in" to contracts with low bids and then slowly ramping up additional costs as the program matured. Legislation tried to keep pace. The 1965 Defense Reorganization Act strengthened McNamara's control, merging service logistics under the Defense Logistics Agency. It aimed to stop nonsense like each branch buying its own radios.

President Johnson, a master of political arm-twisting, played a pivotal role in securing congressional support for Lockheed's C-5A Galaxy transport plane in the mid-1960s. Facing cost overruns and technical problems, the $5 billion program built in Marietta, Georgia, risked cancellation. LBJ, a Texan with deep ties to defense contractors, presented the C-5A as vital for Vietnam logistics and Georgia's economy (where Lockheed employed 60,000 workers). He leaned on his legendary "Johnson Treatment," personally calling and meeting senators like Richard Russell (D-GA), a key Armed Services Committee figure, to ensure funding in the 1965 and 1967 National Defense Authorization Acts. LBJ framed the plane as a national security necessity, downplaying costs while touting jobs. His lobbying countered Air Force doubts and GAO scrutiny, securing $1.3 billion by 1968 despite a $2 billion overrun.

Reform efforts faced cultural hurdles too. McNamara's data-driven zeal alienated generals, who saw PPBS as meddling in their area of expertise. The 1960s were a paradox: bold reforms crashed against stubborn realities. This era is an interesting match to current realities because many of the fundamental technical capabilities developed are still the basic doctrine and platforms that still shape US defense strategy.

◆ ◆ ◆

# The 1970s

The Vietnam War was obviously the driving military and strategic dynamic of the start of the decade. Public trust in government was crumbling after Watergate. Defense acquisition and the industry was already battered by 1960s scandals like the C-5A Galaxy's $2 billion overrun. The decade became a battleground for reform, with Congress, whistleblowers, and Pentagon insiders wrestling to fix a system bleeding cash and credibility. Major events, from the Church Committee's exposé to laws like the International Traffic in Arms Regulations (ITAR), and the Department of Defense Directive 5000, reshaped how the U.S. buys its arsenal. These efforts, driven by themes of transparency, competition, and control, tackled cost overruns and ethical gaps.

The 1970s opened with a hangover from Vietnam's chaos. Defense spending, peaking at $150 billion in 1968, left a trail of bloated contracts and under-performing platforms. Programs like Lockheed's C-5A and General Dynamics' F-111 fighter had embarrassed the Pentagon. Public outrage, fueled by A. Ernest Fitzgerald's *The High Priests of Waste*, demanded accountability. Contractors and Pentagon brass colluded to hide overruns, while Congress, hooked on jobs, looked away. This set the stage for reform, as lawmakers faced pressure to control the military-industrial complex Eisenhower warned about in 1961.

One of the decade's seismic events was the Church Committee. Formally, it was the Senate Select Committee to Study Governmental Operations with Respect to Intelligence Activities, launched in 1975. Chaired by Sen. Frank Church (D-ID), it probed CIA and FBI abuses but also uncovered defense acquisition skeletons. The committee revealed how contractors like Lockheed bribed foreign officials to secure deals. That scandal in Japan cost their Prime Minister his position. These findings, tied to $22 million in Lockheed payments, sparked calls for tighter export controls and ethical oversight.

This push for control birthed the International Traffic in Arms Regulations (ITAR), codified under the 1976 Arms Export Control Act. ITAR wasn't just paperwork; it was a response to the Church Committee's revelations and fears of U.S. weapons fueling global conflicts. The regulations restricted exports of defense technologies, like Boeing's missile guidance systems, requiring State Department approval to prevent tech leaks to adversaries. In 1978 Northrop's ITAR violations for unreported exports resulted in fines and penalties.

Inside the Pentagon, reform took shape with Department of Defense Directive 5000, first issued in 1971 and refined throughout the decade. DOD 5000 was a blueprint for acquisition discipline, mandating a structured process: define requirements, test prototypes, then produce. It aimed to stop disasters like the F-111, where vague specs led to $7 billion in cost growth. The directive pushed milestones like "Key Decision Points" to ensure programs like McDonnell Douglas' F-15 Eagle didn't spiral before approval. By 1975, updates emphasized cost control, requiring "should-cost" analyses to challenge contractor bids.

The 1970 Armed Services Procurement Regulation (precursor to the FAR) got a facelift, tightening rules for cost-plus contracts. The 1974 Congressional Budget and Impoundment Control Act gave Congress more budget oversight, letting it challenge Pentagon requests. The Bradley Fighting Vehicle, starting in 1972, became a cautionary tale. It's $14 billion price tag by 1979, driven by endless design tweaks hilariously shown in the HBO movie *Pentagon Wars,* involved destructive testing and live sheep.

The Church Committee drove other reforms like the 1978 Ethics in Government Act, which was the first limits on the "revolving door" and mandated other disclosures. The 1977 Foreign Corrupt Practices Act barred firms from bribing foreign officials, hitting Northrop with fines for Middle East deals. These laws, while not acquisition-specific, tightened the defense industry's leash.

The decade saw "fly before you buy" testing gain traction, pushed by Sen. William Proxmire, who championed Fitzgerald's cause. The F-16, developed by General Dynamics, was a success story. John Boyd, a maverick Air Force colonel and fighter pilot, spearheaded the development of the F-16 Fighting Falcon in the 1970s as a deliberate counterpoint to the Air Force's push for the costly, complex F-15 Eagle. Boyd, a brilliant tactician known for his "OODA

Loop" (Observe, Orient, Decide, Act), believed the F-15 was over-engineered and not the right platform for the Air Force. In an era of bloated defense budgets post-Vietnam, with programs like the C-5A ballooning to $5 billion, Boyd's "Fighter Mafia" rebelled against the Pentagon's gold-plating culture, advocating for a lightweight, affordable jet.

Boyd's vision, born in the late 1960s at the Pentagon's Systems Command, prioritized maneuverability over bells and whistles. He collaborated with engineers like Pierre Sprey and the General Dynamics team to design a single-engine jet, leveraging insights from his Energy-Maneuverability theory. The F-16, initially the YF-16 proto-type, was pitched as a "low-end" complement to the F-15's "high-end" capabilities, costing $20 billion for 1,388 units. Boyd insisted on a fly-by-wire system, a first for fighters.

The Air Force brass, enamored with the F-15's advanced avionics and twin-engine reliability, resisted. Boyd used backchannels, briefing Congress and allies like Sen. William Proxmire to secure funding through the 1974 Lightweight Fighter Program. By 1979, the F-16 was operational, lauded for its versatility and export success to NATO allies, unlike the pricier F-15.

DOD 5000's structure, ITAR's export controls, and TINA's transparency started to bite. Systemic issues like contractor pricing structures, congressional pork, and Pentagon inertia continued. Vietnam's urgent buys had skewed priorities. The 1970s' legacy is mixed: laws and directives built a framework, but Fitzgerald's "high priests" kept waste alive.

❖ ❖ ❖

# The 1980s

President Reagan's America began pouring billions into defense, determined to outmuscle the Soviet Union. Beneath the bravado, acquisition programs were still collecting headlines for $435 hammers and $640 toilet seats while the the B-1B bomber surpassed $20 billion. The Packard Commission, Reagan's Cold War strategy, and laws like the Competition in Contracting Act reshaped how the Pentagon buys its arsenal.

The 1980s kicked off with a spending boom. Reagan's administration, with the CIA's fresh "missile gap" intelligence, saw the Soviet Union as an existential threat. The Administration more than doubled defense budgets from $134 billion in 1980 to $282 billion by 1987. His approach of massive investment in systems like the Strategic Defense Initiative (SDI) and the Navy's 600-ship fleet was a power play to bankrupt Moscow. Arguably, this strategy should be credited to Andy Marshall, dubbed the "Yoda of the Pentagon." He was a legendary strategist whose quiet brilliance shaped U.S. defense policy for decades. As director of the Pentagon's Office of Net Assessment from 1973 to 2015, he was a master of long-term strategic analysis. In *The Last Warrior* by Andrew Krepinevich and Barry Watts, he's portrayed as a visionary who saw the Cold War not as a sprint but a marathon, advocating a strategy of outspending and out-innovating the Soviet Union to exhaust its economy.

Marshall's analysis, rooted in a net assessment method of comparing U.S. and Soviet strengths, proved the Soviets' centralized system couldn't match U.S. technological and economic dynamism. In the 1970s, he argued for sustained defense investments, like Reagan's $1.6 trillion buildup, including the Strategic Defense Initiative (SDI) and F-16 fighters. His 1980s reports predicted Soviet GDP, strained by military spending (25% of GDP vs. the U.S.'s 6%), would collapse under U.S. innovation in precision weapons and stealth, like the B-2 bomber. Marshall's strategy wasn't about direct confrontation but attrition, forcing the Soviets to chase technologies they couldn't afford. This approach succeeded, and by 1989 the Soviet economy buckled, with defense costs crippling civilian sectors. (Unfortunately, the new Administration ordered the disestablishment of the Pentagon's Office of Net Assessment on March 13, 2025. ONA's civilian employees were reassigned to other "mission-critical positions" within the Department of Defense or dismissed. All ONA contracts were terminated, impacting think tanks and research organizations. The next central hub of strategic planning has not been openly described yet.)

Other weapon system platforms continued to be a mixed collection of cornerstone vehicles and planes still in service as well as dramatic cost overruns. Programs like Rockwell's B-1B, revived after Carter's cancellation, hit $20.4 billion due to rushed designs and cost-plus contracts. Public outrage, fueled by reports of overpriced spare parts, painted the Pentagon as a profligate giant. Reagan's team, while hawkish, knew reform was critical to sustain the buildup.

The Packard Commission was the landmark defense event of 1986. Chaired by David Packard, co-founder of Hewlett-Packard and Nixon's Deputy Defense Secretary, this blue-ribbon panel was Reagan's response to acquisition scandals. The commission's report, *A Quest for Excellence*, diagnosed a system choked by bureaucracy and lax oversight. It found 600,000 requirements in regulations bogging down purchases. Packard's fix was bold: streamline processes, boost competition, and empower program managers. Recommendations included:

- Centralized acquisition authority under a Pentagon undersecretary.
- "Fly before you buy" testing to catch flaws early.
- Multi-year procurement to lock in savings

The commission's ideas weren't new, 1970s DOD Directive 5000 pushed similar rigor, but the Reagan Administration's credibility allowed a stronger approach. By 1987, Congress created the Undersecretary of Defense for Acquisition, streamlining decisions for programs.

Reagan's Cold War strategy amplified reform's urgency. SDI, dubbed "Star Wars," aimed to neutralize Soviet missiles with lasers and satellites, costing $30 billion by 1989. Its complexity demanded tight acquisition controls to avoid C-5A-style overruns. Reagan leaned on private industry, encouraging firms like Lockheed to innovate, but insisted on accountability. His 1983 Executive Order 12352 mandated cost reviews, aligning with Packard's push for "should-cost" analyses. The Navy's A-12 Avenger, a stealth bomber, tested this resolve with its $5 billion overrun, leading to cancellation.

The 1984 Competition in Contracting Act was a cornerstone, mandating open bids for most contracts to curb sole-source deals. CICA is credited with saving $500 million on Boeing's C-17 by forcing competition. The 1986 Goldwater-Nichols Act, while focused on joint operations, boosted acquisition by empowering the Undersecretary. The 1982 Nunn-McCurdy Act, strengthened in 1983, would drag under-performing programs before committee hearings.

Ethical issues like the 1985 Defense Procurement Scandal, Operation Ill Wind exposed bribery and fraud. Contractors like General Dynamics rigged bids, inflating costs for Navy ships by $100 million. The scandal spurred the 1988 Procurement Integrity Act, banning kickbacks and tightening revolving-door rules. Cost control remained elusive as Northrop's B-2 hit $2 billion per plane, despite Nunn-McCurdy's warnings. Multiyear contracts, praised by Packard, saved $1 billion on F/A-18s but failed when specs changed. The decade's rush to outbuild the Soviets with 600 ships and 3,000 planes prioritized speed. Reagan's SDI, while visionary, was a black hole, with untested tech soaking up funds. The new Golden Dome for America concept in 2025 may have a similar outcome.

The decade's reforms were a mixed bag. The Packard Commission, Goldwater-Nichols, and CICA built a framework for efficiency, but scandals and pork dulled their edge. Reagan's Cold War push fueled innovation but fed waste.

❖ ❖ ❖

# The 1990s

The 1990s had the backdrop of the Last Supper consolidation from the end of the Cold War. Defense budgets dropped from $282 billion in 1987 to $260 billion by 1999. The decade sparked a wave of major defense acquisition reform, driven by a simple idea: get smarter with less money. Laws like the Federal Acquisition Streamlining Act and William Perry's Commercial Off-the-Shelf (COTS) memo shook up how the Pentagon bought everything from jets to radios.

The decade kicked off with a jolt with the high visual production value of night vision combat footage of Desert Storm on American televisions. The Navy's A-12 Avenger was cancelled the same year for a $5 billion overrun, showing the Pentagon hadn't shaken the 1960s C-5A curse. A major figure in shaping how the US military is positioned to conduct modern war came from President Clinton's Defense Secretary, William Perry. He was a brainy engineer who saw the Cold War's end as a chance to rethink acquisition. His 1994 report, "Acquisition Reform: A Mandate for Change," argued that the old system was clunky, custom-built, slow, and couldn't keep up with a world of rapid tech and tight wallets. Perry's vision was bold: make defense buying as nimble as Silicon Valley.

His biggest swing was the 1994 COTS memo, "Use of Commercial Products and Practices," a radical call to buy off-the-shelf tech like Intel processors or Microsoft software instead of expensive developmental items. Custom systems, like the B-2's unique radar, took years and billions, while commercial stuff was fast, cheap, and often better. Perry axed thousands of military specifications (milspecs) that were dictating everything from screw sizes to circuit boards. The $40 billion C-17 Globemaster program saved $1 billion by swapping custom avionics for commercial items. This wasn't just a policy tweak; it was a mindset shift, riding the 1990s tech boom.

The 1994 Federal Acquisition Streamlining Act sliced bureaucracy, easing rules for contracts under $100,000 and greenlighting commercial buys. FASA let the Pentagon skip Truth in Negotiations Act audits for off-the-shelf gear. The 1996 Clinger-Cohen Act went further, forcing Pentagon IT buys to mimic commercial practices. These laws were built on the 1980s' Competition in Contracting Act.

The F-22 Raptor, launched in 1991, hit $66 billion by 1999. The Nunn-McCurdy Act, beefed up in 1982, flagged these breaches but rarely stopped programs. Congress, eyeing 20,000 F-22 jobs, kept the cash flowing. The "peace dividend" after the Cold War squeezed budgets, axing projects like the Army's $2 billion Comanche helicopter in 1996.

Dual-use tech, which was civilian technology that could work for military use became a common theme in defense acquisition. This reflected a broader shift in private tech industry R&D investment, catching up and surpassing the DOD's RDTE funding lines. Perry's 1995 Dual-Use Technology Program poured $500 million into projects like flat-panel displays for cockpits.

Reforms leaned on 1980s roots. COTS, FASA, and Clinger-Cohen made buying leaner, but overruns and pork endured.

The 1990s rewired acquisition for a new world, but jobs, lobbying, and red tape are pigs that will never slaughter themselves.

❖ ❖ ❖

# The 2000s

The attack on September 11th and the ensuing Global War on Terror determined the entire approach to defense acquisition throughout the 2000s. Defense budgets skyrocketed from $260 billion in 1999 to $700 billion by 2010, fueling urgent buys for Iraq and Afghanistan. But the acquisition system was out of touch with such an immediate threat, with Defense Secretary Rumsfeld famously stating, "You go to war with the Army you have." The GWOT era shaped the entire military and defense structure for two decades. The major reform efforts like Better Buying Power, the Budget Control Act, and the Gansler Commission mostly tried to address what were seen as major flaws in the system while there were troops in contact.

The unplanned scale of GWOT's demands for body armor, drones, and vehicles exposed the acquisition's sluggishness. Programs like the Navy's Littoral Combat Ship (LCS), launched in 2004, ballooned to $70 billion by 2009. The 1990s' Commercial Off-the-Shelf (COTS) push had not been fully proliferated through the entire defense supply chain.

The Gansler Commission, formed in 2007, was a wake-up call. Chaired by former Pentagon Acquisition chief Jacques Gansler, it tackled the GWOT's contracting chaos in Iraq.

The commission's report, Urgent Reform Required, found 300,000 contractors (more than troops in the country) lacking oversight, with $10 billion in fraud and waste. Cases like Halliburton's inflated $4 billion logistics deals sparked outrage. Gansler's fix: boost the acquisition workforce that was slashed by the 1990s cuts, and tighten contractor rules. By 2008, the Army added 2,000 contract specialists, and the 2009 NDAA mandated fraud audits.

Better Buying Power (BBP) was launched in 2010 by Undersecretary Ashton Carter. Aside from being a bold reform it also inspired the name for this book. BBP wasn't a law but a directive to stretch GWOT's dollars. BBP emphasized competition, saving $2 billion on Boeing's KC-46 tanker by pitting it against Airbus. It also revived the 1990s' Other Transactions Authority (OTA), letting firms like SpaceX bid on $500 million satellite launches.

Congress threw its weight behind reform with the 2011 Budget Control Act (BCA). The BCA capped defense spending, triggering $500 billion in cuts via sequestration. This killed the Army's massive $40 billion Future Combat Systems program. Sequestration was a blunt axe, delaying F-35 buys and cutting 10,000 jobs. The 2009 Weapon Systems Acquisition Reform Act (WSARA) complemented BCA, mandating independent cost estimates for programs. WSARA's Nunn-McCurdy tweaks flagged LCS breaches, but Congress, tied to 4,000 shipyard jobs, kept it afloat.

The MRAP program was a shining exception. Launched in 2006 to counter Iraq's roadside bombs that were killing 1,000 soldiers a year. MRAPs were built by firms like Oshkosh and BAE, delivering 27,000 vehicles by 2012. (This fielded capability saved thousands of lives, including my own, when my vehicle was hit by a pressure plate IED in 2009.) Rapid prototyping and competitive bids kept costs at $1.5 million per unit, a stark contrast to the F-22's $350 million per plane. MRAP's success, driven by GWOT's urgency, showed reform's potential when stakes were visceral.

It is impossible to address the defense enterprise during the GWOT era without referencing one of the most competent and honorable Senators of the modern era. Senator John McCain, a Vietnam War hero and longtime Arizona Republican, was a relentless advocate for defense acquisition reform during the Global War on Terror. As a senior member and later chairman of the Senate Armed Services Committee, McCain targeted the Pentagon's wasteful, slow, and often corrupt acquisition system, which he saw as a national security crisis.

McCain's crusade began with high-profile scandals. In 2003, he spearheaded investigations into Boeing's $23 billion KC-46 tanker lease deal, exposing ethical violations resulting in jail time and costing the Boeing CEO his job. McCain's dogged oversight forced a competitive rebid in 2008, saving billions by pitting Boeing against Airbus. He argued GWOT's pace demanded efficiency, not waste like the 1980s. His 2007 Defense Acquisition Reform Act (S. 32) tackled "gold-plating," where unneeded features increased costs.

As SASC chairman in 2015, McCain drove sweeping reforms via the 2016 National NDAA that are still in place now. The NDAA empowered service chiefs with milestone decision authority, tying accountability to results, and expanded rapid acquisition paths, like Other Transactions Authority. McCain, inspired by the 1986 Packard Commission, pushed "fly before you buy" testing to catch flaws early, as seen in the Littoral Combat Ship's $70 billion overrun. He also championed the 2009 Weapon Systems Acquisition Reform Act, mandating cost audits that flagged F-35 breaches, though its 180,000 jobs kept Congress lenient.

# CHAPTER 12

# An Imperfect Fifth Estate

*"This would be a great time in the world for some man to come along that knew something."*
—Will Rogers

FROM THE INTRODUCTION, THE THEME and central tenet of this book has been that the defense industry is the real "hard mode." In February 2025, Anduril Industries, a defense tech startup known for autonomous drones and AI systems, launched a bold recruitment campaign called "Don't Work at Anduril." Unlike typical corporate hiring ads, it used reverse psychology to attract mission-driven talent, targeting cities like Boston, Atlanta, and Seattle, where young engineers and tech professionals thrive. The campaign's edgy, in-your-face style made it go viral, sparking buzz on social media and boosting job applications by over 30 percent.

The campaign's centerpiece was gritty, graffiti-style ads plastered across bus stops, subway stations, and billboards near MIT, Harvard, and Georgia Tech. The text read "Work at Anduril.com," with "Don't" spray-painted over it in red, white, or blue, mimicking vandalism. A short online video leaned into the sarcasm, featuring a fictional employee griping about the job's intensity: long hours, field tests at dusty ranges, and demands for "Pacific Rim-sized robots." Founder Palmer Luckey even popped up in his signature Hawaiian shirt, poking fun at the chaos. The message was clear: Anduril's work is grueling, in-person, and tied to supporting U.S. troops, perfect for relentless builders seeking purpose.

On LinkedIn, recruiters swapped profile banners for purple "Don't Work at Anduril" graphics, doubling down on the vibe: "Hard work, hard problems, hard mode." Posts on X teased the campaign's intensity, framing complaints like "no swanky hotels" as badges of honor for those who'd "run through walls" for the mission. The campaign screamed patriotism, with American flags and a call to transform military capabilities, setting Anduril apart from cushy tech giants.

It worked because it was authentic. Anduril didn't sugarcoat its culture, it challenged people to self-select out if they don't align. By targeting those who crave purpose over perks, it drew thousands to its open roles, from AI coders to drone designers.

※ ※ ※

## Corporate Power

With almost a trillion dollars in internal spend and over a hundred billion in arms exports, the industry doesn't just build hardware; it extends U.S. hegemony, absorbs economic risks, and provides a "hard" insurance policy for GDP growth. Economically, the industry acts like a shock absorber, which gives Congress one of its most direct paths to employment and distribution of the national budget, with its 3 million jobs anchoring communities like Fort Worth or Bath, Maine. In 2010, $150 billion in exports fueled growth. The industry's risk-taking innovation has returned dividends via dual-use tech like GPS.

Picture a Virginia shipyard at dawn, where welders' sparks light up the hull of a new destroyer, each arc a testament to a nation's resolve. This is the U.S. defense industry, not just factories and contracts, but a fifth estate. A force wielding America's might while shouldering significant risks. This is a call to arms, not for war, but to champion an industry that marries purpose with power, innovation with grit. Through Shyam Sankar's "First Breakfast," Alex Karp's *Technological Republic*, and the specter of an isolationist U.S., we should rally behind a sector that, despite flaws, holds the line for freedom and stability in 2025.

$150 billion in arms exports serves as an investment in the existing international order. While critics describe this as a war-mongering nouveau imperialism, it is very hard to find a credible description of what geopolitics would look like with China, Iran, or Russia expanding their international influence.

Shyam Sankar's "First Breakfast," is a clarion call to shake up this giant. Sankar, the Palantir CTO, sees the Pentagon's monopsony as a creativity killer. His 2024 theses demand a defense tech renaissance, integrating startups to counter the five-prime stranglehold. Think of the MRAP program: $50 billion, 27,000 vehicles in six years, saving countless lives. Sankar wants that speed for drones and AI, using OTAs and Commercial Solutions Openings to actually grow the defense industrial base. Thomas Schelling's 1966 *Arms and Influence* underscores the need for agile systems to deter foes proving that not every concept is new.

Dr. Alex Karp's *Technological Republic* takes the baton, envisioning a defense industry powered by Silicon Valley's spark. Palantir's CEO sees AI and software as 2025's battleground. Kenneth Waltz's *Man, the State, and War* from back in 1959 was already framing this as a state-driven race, where tech secures sovereignty. Karp's republic isn't utopian; it's a pragmatic bet on brains over brawn, vital for a cyber-driven future.

But what happens if America turns inward? An isolationist U.S. upending traditional NATO structures would leave a global vacuum. China's open threats to invade Taiwan, Russia's "annexation" of Ukraine, and Iran's willingness to launch attacks in the Middle East prove that the era of Pax Americana is over. CSIS published a report in 2023 that seems to have described the international script for the next few years: Russian military resurgence, Iran's drones in Yemen, and China's South China Sea expansion. Deterrence hinges on the presence of US military power, and the value of asymmetric technological superiority is even more important as threat powers activate.

❖ ❖ ❖

## In Our New Trump Era

It is impossible to overstate the bold attempts the new Trump Administration is making to completely reshape the

DOD and the defense industrial base. The defense industry is trying to keep up with real-time modification of fundamental US policy. Secretary Hegseth is a unique outsider that does not come from a background in the industry or from government. The Administration is shaking up the defense world with a flurry of Executive Orders aimed at overhauling acquisition, reenergizing the defense industry, and rewriting the rules of federal procurement. At the time of publication, it is still difficult to estimate what the 2025 or 2026 defense budgets will be. There are competing narratives from an early DOGE plan to cut 8% across the DOD and then a pivot a month later to a $1 trillion budget plan. Secretary Hegseth definitely brings a bold, outsider's zeal to reshape a military he sees as bogged down by bureaucracy and waste. His reforms target speed, innovation, lethality, and either cost-cutting or dramatic budget expansion.

The centerpiece is Trump's April 9, 2025, EO, "Modernizing Defense Acquisitions and Spurring Innovation in the Defense Industrial Base," a sweeping directive to revamp how the Pentagon buys everything from jets to software. Hegseth, tasked with delivering a reform plan by June 8, 2025, is told to prioritize commercial solutions like off-the-shelf tech and lean on Other Transaction Authority (OTA) agreements to bypass FAR rules. The EO demands a review of all 72 Major Defense Acquisition Programs (MDAPs) within 90 days, flagging any over 15% behind schedule or budget for cancellation, like the Navy's nine delayed ship programs or the Air Force One replacement which is five years late. The EO also pushes a "ten-for-one" rule, slashing ten regulations for each new one. This could trim the 1,500-page Defense Federal Acquisition Regulation Supplement

(DFARS), but while there is merit to starting over, many of the clauses included are the results of hard lessons learned.

This isn't just about paperwork. Hegseth's vision, as he told troops at the Army War College on April 23, 2025, is to ditch "PowerPoint bloat" and funnel savings into "mission-critical" gear like drones and cyber tools. He's eyeing a leaner acquisition workforce, with a 120-day plan to "right-size" with reports of plans to cut over 20,000 DOD civilians, and dramatically cut the General Officer billets. The Administration and the Defense Secretary are definitely conducting a live experiment into the scope of their authority to reshape the department.

Efforts to recreate the defense industry are equally audacious. Trump's April 9 EO pushes commercial tech, but other policy descriptions have embraced a "right to repair" mandate that may invalidate the IP protections central to dual tech. Hegseth's March 6, 2025, software memo made Commercial Solutions Openings (CSOs) and OTAs the default for software, speeding up buys like Palantir's $500 million AI contract. A separate April 9 EO, "Reforming Foreign Military Sales," streamlines exports, cutting International Traffic in Arms Regulations (ITAR) red tape to boost firms like Boeing's F-15 sales to allies. A third EO, on maritime industries, invokes the Defense Production Act to revive shipyards, targeting Virginia's 10,000 jobs and countering China's naval edge.

Rewriting the FAR is the boldest stroke. President Trump's April 15, 2025 EO "Restoring Common Sense to Federal Procurement," orders the FAR Council to scrap any non-statutory rules within 180 days, targeting "anti-competitive barriers." The FAR is a 2,000-page behemoth that governs $600 billion in federal buys, but with a hard reset, there may be a period of the Wild West where contracting officers and defense contractors don't know what is feasible. The April 16 EO, "Ensuring Commercial, Cost-Effective Solutions," mandates market research to favor commercial goods building on 1994's FASA.

Secretary Hegseth's military overhaul is a wild card for the defense industry. His February 19, 2025, memo ordered an 8% budget cut yearly, aiming to save $50 billion by 2026 for priorities like the southern border, but budget planning discussions are increasing the topline dramatically. Requiring consulting contractors to defend their programs is a unique approach, considering they bid their scope and costs in response to Agency requests. Finally, the defense industry is trying to react to daily changes in foreign policy, like Ukraine and Yemen, as well as tariff policy that does not seem to adequately account for international trade agreements fundamental to defense industrial base supply chains.

# Conclusion

THE DEFENSE INDUSTRY IS ONE of the rarest sectors, it connects welders and blue-collar machine workers to the software engineers in Palo Alto. This is the U.S. defense industry, not just steel and code, but a fifth estate, a force pulsing with America's spirit. It's a call to arms, not for battle, but to rally behind a mission that guards the ideals of Western Civilization. With hard technology and the chaotic oversight of 535 congressional "board members," it's a challenge worthy of our nation's brightest.

Picture a world where this industry falters. Imagine Pacific allies facing a rising power's navy, unchecked, their ports shadowed by foreign fleets. Envision Eastern Europe, abandoned, where old empires reclaim lost ground, or the Middle East, swarmed by drones no one counters. This isn't abstract; it's the weight of history. Decades ago, factories churned out thousands of planes, smashing tyranny's grip. In recent wars, armored vehicles rolled out in a sprint, saving soldiers from roadside ambushes. These triumphs show the industry's soul building not just weapons, but stability. Our best minds must ensure this legacy endures.

Shipyards and labs anchor towns, breathing life into places where factories might otherwise rust. It is an area where re-shoring and US manufacturing capability are necessary. It's a risk-taker, pouring genius into tech that sparks civilian breakthroughs. The technological challenge is relentless. Defense tech demands brilliance under pressure.

Then there's Congress, a 535-strong board of directors, each with a district's pulse in their votes. They steer the industry's path, balancing hometown jobs with national needs. A senator from Ohio fights for a tank plant, not just for steel, but for families who'd lose everything if it closed. This isn't mere politics; it's the human side of power, where every decision ripples to a factory town or a coastal yard. Navigating this maze, where every vote pulls a different string, should be debated in the open.

This mission stirs the soul. Those armored vehicles, rushed to warzones, didn't just roll; they carried hope, shielding troops from harm. Every plane, every line of code, bears the weight of lives protected, nations secured. It's the hardest arena, where purpose fuses with the dual challenge of trying to remain economically viable while also creating true military capacity.

And here's the closing truth: the U.S. defense industry is *Buying Power* for America every day. It doesn't just craft weapons; it buys a future. Every ship welded, every algorithm trained, invests in a world where America leads with resolve. Facing Congress's 535 voices and technology's frontier, it demands our brightest: thinkers, builders, dreamers, to maintain a fifth estate that stands tall. The American experiment is not a Norman Rockwell painting, and perfect is never on the menu. The industry mirrors the national identity: a constantly shifting collision of money, technology, and global geopolitics. This industry is one of the largest players in the game theory going on around us. The defense industry, for all its faults, provides the credibility and authority to our nation.

# References

Babbitt, Joel D. *Half Price Negotiating*. 2022.

Barnard, Chester I. *The Functions of the Executive*. Harvard University Press. 1968.

Blattman, Christopher. *Why We Fight*. Penguin Random House. 2022.

Brose, Christian. *The Kill Chain*. Grand Central Publishing. 2020.

Burton, James G. *The Pentagon Wars*. Naval Institute Press. 1993.

Bussolini, Jake. *The Last Chapter*. AuthorHouse. 2014.

Cibinc, John; Nash, Ralph; Yukins, Christopher R. *Formation of Government Contracts*. Wolters Kluwer. 2011.

Clarke, Richard A; Knake, Robert K. *The Fifth Domain*. Penguin Random House. 2019.

Cropsey, Seth. *Seablindness*. Encounter Books. 2017.

Dalio, Ray. *The Changing World Order*. Simon and Schuster. 2021.

Dole, Bob. *Great Political Wit*. Doubleday. 1998.

Fox, Ronald J. *Arming America*. Harvard University Press. 1974.

Fox, Ronald J. *The Defense Management Challenge, Weapons Acquisition*. Harvard Business School Press. 1988.

Gabor, Andrea. *The Capitalist Philosophers*. Random House. 2000.

Gansler, Jacques S. *The Defense Industry*. MIT Press. 1980.

Gryta, Thomas; Mann, Ted. *Lights Out: Pride, Delusion, and the Fall of General Electric*. Harper Collins. 2020.

Hart, Amber; Lisa Shea Mundt. *BD 360*. The Pulse of GovCon. 2024.

Hartung, William D. *And Weapons for All*. Harper Collins. 1994.

Hartung, William D. *How Much Are You Making on the War, Daddy?* Nation Books. 2003.

Hartung, William D. *Prophets of War*. Nation Books. 2011.

Fitzgerald, A. Ernest. *The High Priests of Waste*. Norton. 1972.

Fitzgerald, A. Ernest. *The Pentagonists*. Houghton Mifflin. 1989.

Janik, Richard. *Fires in the Sky*. 2024.

Jenkins, Dale A. *Diplomats & Admirals*. Aubrey Publishing. 2022.

Jones, Wilbur D. *Arming the Eagle*. Defense Systems Management College. 1999.

Kapstein, Ethan Barnaby. *The Political Economy of National Security*. McGraw-Hill. 1992.

Klein, Maury. *A Call to Arms: Mobilizing America for World War II*. Bloomsbury Press. 2013.

Koprince, Steven J. *The Small-Business Guide to Government Contracts*. American Management Association. 2012.

Krepinevich, Andrew F; Watts, Barry. *The Last Warrior*. Basic Books. 2015.

Long, Kim. *The Almanac of Political Corruption, Scandals, & Dirty Politics*. Bantam Dell. 2007.

Macgregor, Douglas A. *Breaking the Phalanx*. Center for Strategic and International Studies. 1997.

Macgregor, Douglas A. *Transformation Under Fire: Revolutionizing How America Fights*. Praeger. 2003.

Melman, Seymour. Pentagon Capitalism: *The Political Economy of War*. McGraw-Hill. 1970.

Miller, Chris. *Chip War*. Scribner. 2022.

Milnarchik, Christoph. *Government Contracts in Plain English*. 2019.

*On Strategy*. Harvard Business Review Press. 2011.

Pavelec, Sterling Michael. *The Military Industrial Complex and American Society*. ABC-CLIO. 2010.

Perkins, John. *The New Confessions of an Economic Hitman*. Berrett-Koehler. 2016.

Rothbard, Murray N. *Anatomy of the State*. BN Publishing. 1974.

Sampson, Anthony. *The Arms Bazaar in the Nineties*. Hodder and Stoughton. 1977.

Scharre, Paul. *Army of None: Autonomous Weapons and the Future of War*. Norton. 2018.

Schelling, Thomas C. *Arms and Influence*. Yale University. 1966.

Segel, Kenneth R. *The Government Subcontractor's Guide to Terms and Conditions*. Management Concepts. 2009.

Shafritz, Jay M. *Words on War*. Prentice Hall. 1990.

Tsouras, Peter G. *The Greenhill Dictionary of Military Quotations.* Greenhill Books. 2000.

Waltz, Kenneth N. *Man, the State and War: A Theoretical Analysis.* Columbia University Press. 1954.